CRÆFT

Dr Alexander Langlands is a British archaeologist and historian. He is a lecturer in medieval history at Swansea University and a regular presenter for the BBC and Channel 4. His current and ongoing TV projects include *Full Steam Ahead* (BBC Two) and *Britain at Low Tide* (Channel 4). He lives in Wales with his wife, two children, chickens, and bees.

CRÆFT

How Traditional Crafts Are
About More Than Just Making

ALEXANDER LANGLANDS

FABER & FABER

First published in 2017
by Faber & Faber Ltd
Bloomsbury House
74–77 Great Russell Street
London WC1B 3DA

Typeset by Faber & Faber Ltd
Printed and bound in the UK by CPI Group (UK), Croydon CR0 4YY

Gardening catalogue images on p. 35 digitised by the Internet Archive

A CIP record for this book
is available from the British Library

ISBN 978–0–571–32440–8

FSC
www.fsc.org
MIX
Paper from
responsible sources
FSC® C020471

2 4 6 8 10 9 7 5 3 1

For Nicola

CONTENTS

FOREWORD

THE STRIMMER AND THE SCYTHE

I REMEMBER THE FIRST time I used a scythe. I was in my mid-twenties and, sick of city life, I had taken up the tenancy of a cottage in the middle of nowhere. Neglected for years, the garden was a waste ground of invasive weeds, molehills and tumbled-down fences. At first I relished the challenge of taming this wilderness, but after just one year the burden of mowing the grass, strimming the various rough patches and trimming the hedges began to grind me down.

Perhaps the most irritating part of the process was the maintenance of the various petrol-powered cutting machines I had to use. The strimmer, a length of steel tube with an engine at one end and a head of rotating nylon cord at the other, was a particularly truculent contraption. I always dreaded trying to get the blasted thing started after its long winter hibernation. Ensuring the mix of oil and petrol was just right for the highly strung engine, clearing the air filter, decarbonising the spark plug, replenishing the nylon cord, oiling the head and flushing the carburettor were all jobs that had to be undertaken before endlessly yanking away at the starter cord, desperately hoping it would fire into life. And when it finally roused itself from its winter slumber and revved up to fever pitch, on went the protective gear: steel toecaps, goggles, gloves and ear-defenders. I would submerge myself into a day of monotonous buzzing and rattling, flaying the emerging spring vegetation.

However, one morning in late April, no matter what I did I just could not get the damned thing to start. In the days of my youth I would have taken it to my father – a man of the post-war generation when everyone was a hobbyist mechanic – and with a bit of tinkering he would soon have had it going. But I was on my own here. In the middle of nowhere. And no amount of swearing and cursing was going to improve the situation. Dismayed, as I cast my eye over the unruly undergrowth encroaching on the last vestiges of lawn around the cottage, my mind recalled an implement I had hanging up in the back shed. It was a scythe I'd purchased a few years before at a car boot sale, for the princely sum of ten pounds. Carrying it back to the car that day, I had conjured up a romantic vision of myself emulating the farmhands of Old England, slashing through acres of luscious meadow grass between manly swigs from a cider flagon. Blunt and rickety scythe in hand, I set about cutting and quickly developed a pendulum-like hacking motion. Progress was slow, but it was working – and I was living the dream.

It was lucky for me that one of the older gamekeepers caught sight of me. As he pulled up in his clapped-out Land Rover he leaned out of the window and laughed. 'I can see you've never used a scythe before, boy.' Within seconds he was smearing a drop of spit down the blade with a whetstone, working up a fine abrasive paste and softly grinding a shining edge on the black patina of the antiquated iron. Razor sharpness was everything. And the technique he demonstrated was different too. Holding the blade parallel to the ground and as far away from the body as was comfortable, he drew it towards himself in an arcing motion, slicing – not hacking – through the undergrowth. The hollow ringing sound of the blade scything through the grass and weeds was clean and appealing. But what's more, the speed and effectiveness was astonishing. On the back swing a brushing technique could be adopted with the rear of the blade, teeing up any fallen plants to be sliced through on the returning swipe. I was impressed. And while the job had probably taken me a fraction longer than with a strimmer, I'd enjoyed listening to the sound of the birds while I worked.

That summer the scythe became the tool of choice. Relieved of the rigmarole of fuelling, servicing and maintaining the strimmer, scything could be conducted on a whim, the scythe plucked from the toolshed and employed for an hour or two here and there. My technique improved. I became stronger and began to feel less exhausted at the end of a stint, and almost matched the time taken to do the same job with a strimmer. And the shape of the garden changed too; straight lines gave way to sweeping curves and corners became rounded. Scythed twice that year, the variable stubble of my small meadow created an attractive environment for a variety of grasses and wild flowers, which in turn supported a host of different insects. As autumn reached for her golden crown, I realised that I'd taken a traditional way of doing something and had found that, on my terms, it was just as effective as the mechanically charged, petrol-powered methods of today.

And so, my relationship with *cræft* had begun.

PREFACE

WHY CRÆFT?

I GUESS I NEEDED to make a distinction between how we think about the modern definition of craft and what it meant when it first appeared in the English language over a thousand years ago. In a keynote lecture given to the Heritage Crafts Association in 2013, Sir Christopher Frayling echoed the sentiments of David Pye, in *The Nature and Art of Workmanship*, when he called craft, 'a word to start an argument with'. I don't want to start any arguments but it's true: craft has become so ubiquitous that it's increasingly difficult to state with any exactitude a definition precise enough to satisfy everyone. Certainly, it has something to do with making – and making with a perceived authenticity: by hand, with love; from raw, natural materials; to a desired standard. It doesn't necessarily have to result in an object, though. A recent craze for craft beers means that we can consume craft and essentially come away with nothing to show for our purchase – except perhaps a slightly fuzzy head the next day. In the world of art it can be a methodological process as much as a conceptual tool. In the world of luxury, a reassurance that you are acquiring the very best product money can buy. In the world of the everyday, the success of the retail giant Hobbycraft is the best illustration that we still revel in the pastime of using our hands to make something that can be given, enjoyed and cherished.

But even in today's versatile use of the word craft there is only the faintest overlap with the definition *cræft* had when it first appeared in written English over a thousand years ago. The *Oxford English Dictionary* can find no one word to exchange, like for like, for Old English *cræft*, and instead offers an amalgam of 'knowledge, power, skill', and an extended definition where a sense of 'wisdom' and 'resourcefulness' surpass in importance the notion of 'physical skill'. It would seem that we can't quite put our finger on exactly what *cræft* was.

It is this inability to assign a precise contemporary meaning that justifies the ideas put forward in this book of a *lost* knowledge and of how traditional crafts, as we know them, are about so much more than just making. We don't have *cræft* in our lives any more. Our Anglo-Saxon ancestors certainly had it, but at some point we mislaid it and with it its true meaning. Over the course of the last fifteen years I've found many occasions to think through the idea of a lost knowledge. As an archaeologist I'm constantly confronted with the material culture of past societies: objects that were once fashioned, used, altered and discarded. Through the analysis of these objects, archaeologists attempt to draw conclusions about the nature of the human condition and, in particular, how our thinking, our actions and our relationship with our environment have changed over time.

I rarely study anything archaeological that is more recent than the fifteenth century. But for a period of ten years, from 2003 to 2013, I participated in a number of television series for the BBC that charged our various team members with recreating life as it would have been on British farms from the seventeenth to the twentieth centuries. Unlike the work of an archaeologist, whose task it is to survey essentially static remains, in the making of *Tales from the Green Valley* (2005), *Victorian Farm* (2008), *Edwardian Farm* (2010) and *Wartime Farm* (2012), I often observed material processes in action, and was involved in how an archaeological record was actually created. In filming these overtly nostalgic historical programmes, I was consistently confronted with a narrative of the old ways, a sense of unrelenting change and a feeling that something was for ever being lost.

At first, I railed against this cliché of retrospective regret. For the angry young man that I was back then, Billy Bragg's invocation to damn nostalgia as the 'opium of the age' rang loudly in my ears. But gradually I began to realise there was more than a kernel of truth in the nostalgic motifs we were revisiting.

Society, I concluded, *was* losing something. As I became more and more engrossed in the traditional ways – and not just historical methods of farming but ways of making and living in the past – it occurred to me that the modern world was depriving us of many of these skills. What I saw as a wider knowledge – one that enabled us to exist in a world where our sustenance and survival depended on our interactions with the materials we had at our disposal – was slowly slipping from our grasp.

Having finally got myself up to speed with the digital world, I begin to wonder whether the vast complexity and infinite interactions digital technology promises are in fact doing quite the opposite: are they actually narrowing our sensory experiences? We're increasingly constrained by computers and a pixelated abridgement of reality that serves only to make us blind to the truly infinite complexity of the natural world. Most critically, our physical movements have been almost entirely removed as a factor in our own existence. Now all we seem to do is press buttons.

Richard Sennett, in his 'template for living' *The Craftsman*, talks about craftsmanship as the state of being engaged: how we interact materially, with each other and our immediate surroundings. Perhaps we should consider this as a key component of the long-lost *cræft*. Against a narrative of progressive technological innovation, what has happened to *cræft*, the indefinable intelligence of our Anglo-Saxon forebears? What reasons lie behind its drift into obscurity? Chiefly, I accuse industrialisation and the introduction of cheap and vastly superior forms of power – resulting in what I call our illiteracy of power. We simply don't need to factor power into how we make from and process raw materials. Nowadays, with a flick of a switch, we can generate what would take far more time, human energy and cost to produce by hand. The point when industrial processes emerged as the dominant means of production was the point at which the concept of craft as a form of art emerged –

as a self-conscious counterpoint to factory-made goods. Craft became defined in opposition to industrial manufacture.

Mechanisation too, and especially the small electrical motor, has largely robbed us of the need to be physically skilful and dextrous. Everyday skills, such as mixing ingredients with our hands, have been given over to electrically driven implements. The growth of formal knowledge – an intellectualised understanding of the world – has meant that learning through practice, by rote and experience, has been relegated. It's more customary today to refer to the text – the formal knowledge – of the manual than it is to take something apart and see how it actually works.

I'm not saying that either of these developments is necessarily bad. There are many occasions when I probably should have consulted the manual before taking a malfunctioning machine apart. But mechanisation has changed the way we think, the way we build knowledge; so familiar has post-industrial power become that we genuinely find it hard to relate to the world before it. This may be why a true definition of *cræft* is so remote to us: we have forgotten how to think like the generations before the Industrial Revolution.

FOR ANYONE WHO chooses to read this book aloud, *æ* is a diphthong: a complex speech sound beginning with one vowel and gliding into another. In classical and modern languages it's pronounced using a range of sounds, but in the Old English alphabet, where it's known as an 'aesc' or 'ash', its pronunciation falls somewhere between 'a' and 'e', and in the case of *cræft*, I usually say a word that sounds more like creft than craft. But that's just me. I'm from the south-east of England and speak with an estuary accent: far from posh, a step up from cockney.

1

DEFINING *CRÆFT*

IN THE LATER years of his life, Alfred the Great, celebrated king of the English, defender of the Church and scourge of the Vikings, settled down to prove that his pen could indeed be as mighty as his sword. He was a pious king. Years of being taken to the brink of defeat had caused him to consider that his fate was being determined less by his own guile and more by some higher and mightier power. This soul-searching in the twilight of his life delivered him to the works of wise men – churchmen and philosophers – and as he read and considered their writings, the paternalistic king took it upon himself to spread the word. He set about translating documents, for the first time, from Latin into Old English, embellishing his translations with his thoughts and ponderings, and in the process providing us with a tantalising glimpse into the mind of one of England's most eminent statesmen.

The youngest of five brothers, Alfred had to sit back and watch as the brutal early medieval world took its toll on his kinsmen. His brothers, Æthelbald, Æthebert and Æthelred I, had between them reigned for little more than fifteen years during a period that saw the fledgling kingdom of England driven to its very knees and taken, as one commentator put it, into the 'crucible of defeat'. Northumbria had fallen, Mercia was in its final death throes and Wessex stood alone, teetering on the verge of collapse as a Viking army of unprecedented proportions ravaged the nation's people. Almost inevitably, Alfred found himself next in line for a throne that promised nothing but unrelenting warfare, unavoidable slaughter and the impossible task of rescuing the Anglo-Saxon world from the prospect of eternal oblivion.

But he did it, and in the achievements of his age, Alfred's talent clearly extended beyond mere rabble-rousing, good sword technique and the ability to stand fast in a shield wall beset by the trancelike psychotic rage of the Viking berserkers. While these prerequisites undoubtedly allowed Alfred to secure power, it was with ideological, political, economic, administrative and strategic tools that he maintained the status quo and even, for a fairly lengthy time, retained peace in the land.

His taming of Guthrum, through his enforced conversion to Christianity after the Battle of Edington, brought the Scandinavian king into a wider European cultural milieu that, through co-operation rather than violence, would reap wealth and rewards for both parties with significantly less loss of human life. In political terms, the boundary known as the Danelaw, an arbitrary line drawn diagonally across England from the east of London to the Wirral in the north-west, gave Guthrum's followers the opportunity to settle and exchange the battleaxe for the plough.

Alfred's programme of fortified town construction throughout the kingdom of Wessex was designed as much with a view to consolidation of his power as it was to establish the economic future of the kingdom. And his ambitious plan to have every child of noble birth schooled in the English language paved the way for an administrative and legal foundation that would set in motion the emergence of the English state machinery that William the Conqueror was so keen to wrest control of in 1066. Putting divine intervention aside, it was clear that Alfred was a resourceful chap. He realised that the package of kingship involved a greater range of power, skill and wisdom than your average 'Dark Age', mead-swigging, skull-cracking warlord.

It is through Alfred's writings, and especially his translations of ancient texts, that we can enter his thought patterns and gain an insight into how he perceived his own talents and those he recognised in other people. One word in particular crops up over and over again as the warrior-turned-scribe wrestled to find a lexical range in the Old English tongue to interpret what he was confronted with in the Latin texts of classical writers. That word is *crœft*.

In fact, so frequent is the word's appearance in Alfred's work, it might be considered nothing more than a catch-all term employed as a consequence of Old English's inability to match Latin's diverse wordplay. But only a fool would accuse the Anglo-Saxons of being anything other than the finest wordsmiths. Throughout all his translations and prose, the specific contexts within which *crœft* is used show that Alfred was grappling not just to replace, sense for sense, but to describe a quality or state of being; an almost indefinable knowledge or wisdom.

I'm not saying that *crœft* didn't mean to the Anglo-Saxons what it means to us today: a physical skill, ability or dexterity. But of the 1,331 appearances of the term in all Anglo-Saxon documents, whether used singly or as a compound, in the greatest number of cases the meaning is of power or skill in the context of knowledge, ability and a kind of learning. Furthermore, a sense of mental skill – merit, talent or excellence – occurs as many times as the sense of mere physical skill. But Alfred does something else with *crœft* in his translation of one particular text, Boethius's *Consolation of Philosophy*. On a significant number of occasions he uses *crœft* to translate the Latin *virtus*, meaning virtue, in the sense of spiritual skill or excellence. In a study of Alfred's debt to vernacular poetry, the historian Peter Clemoes writes that Alfred's uses of *crœft* are best explained as 'the organising principle of the individual's capacity to follow a moral and mental life'. Alfred 'The Great' becomes Alfred 'The Life Coach' some twelve hundred years before the publication of Richard Sennett's so-called 'template for living', *The Craftsman*.

Over the following millennium the word *crœft* has a rich variety of meanings. Alfred praised God for the *Wundorlice crœfte* (wonderful craft) with which he had shaped the earth, and this sense survives in biblical tracts right up to the sixteenth century. In *Game and playe of the chesse*, printed by William Caxton in 1474, for example, a 'romayn' (Roman) would choose to defeat their opponent less by use of 'subtilnes' (subtlety) and more by an overt 'craft and strengthe of armes'. This powerful and almost brute force seems a long way from the form of intelligence seen in other contexts. The 'poetrie' of Gavin Douglas's 1513 translation of Virgil's *Æneid* is conducted in a craft-like fashion, and 'love', in Chaucer's 1381 *Parliament of Fowls*, is treated almost as a profession for which life is too short for a 'craft so long to lerne'. Clearly, craft in these more amorous settings is anything but a natural gift and rather something that must be studied and learned in depth.

This ties in with our current understanding of a craft as something that one must take time to learn in order to be a competent let alone master practitioner. Tutoring begins with an apprenticeship, and a consistent association between craft and making as a vocation can be observed from its earliest mention all the way to the present day. William Langland in *Piers Plowman* (1362) wrote of 'Taillours, tanneris and masons' among 'mony other craftes'. In 1758, on the eve of the Industrial Revolution, Samuel Johnson in his series of essays entitled *The Idler* talks of the crafts of 'Shoe-maker, Tin-man, Plumber, and Potter'. Of that special something, however, that special quality or skill, Joseph Moxon's *The Doctrine of Handy-works* (1678) is most explicit when he marvels at the joiner in his 'craft of bearing his hand so curiously even, the whole length of a long board'.

But very early on in the history of this word we start to see something happen that casts craft in a much more negative light. The demonic side is there from the beginning and is found in craft's most famous compound: witchcraft. In the Anglo-Saxon period *wiccecræfte* is glossed as the Latin for *necromantia* (necromancy, communicating with the dead), and *demonum invocatio* (the calling up of demons). In the later medieval period, although less deviant, in John Trevisa's English translation of the Benedictine monk Ranulph Higden's *Polychronicon* (1387), we are told that, of some men, we must be wary more of their guile than of their craft. This suggests that the two have not yet overlapped and that the latter is certainly not as bad as the former. Yet the mere association with guile is enough to taint. By the time Thomas Hobbes was publishing *Leviathan* in 1651, craft is explicitly linked to a 'Crooked Wisdom', and in 1856, in the American poet Ralph Waldo Emerson's commentary on *English Traits*, there is a clear association between actions of poisoning, way-laying, assassination and 'craft and subtlety'. How did this come about?

The problem is, of course, that craft is as much a loaded term as it is a pragmatic description of how one earns one's livelihood. In its defamatory use value judgements are being made. And this is the issue with simply exploring the history of a word. To uncritically accept the present reading of craft is to fail to see beyond its recorded history to the actual ability and skill it purports to describe. From the very beginning the use of the term witchcraft was heavily laden with a Christian bias, a worldview within which any other form of belief system was seen as heretical. We've come a long way since the medieval ducking pond and the witch-burnings of the seventeenth century and are much more accepting of different religious practices. With its emphasis on natural cures, remedies and spiritual well-being, modern witchcraft is regarded as a positive thing – the antidote to our over-medicalised world. And more than a thousand years ago witchcraft represented an intelligent set of ritual practices.

Today, it is the word crafty that has suffered most. Although consistently used to describe someone wise, ingenious, clever, dextrous and skilful, from its first usage right up until the nineteenth century, the negativity it's associated with today is well illustrated in Shakespeare's *King John*, where a love that is 'craftie' is a love that is 'cunning'. More explicitly, Hobbes, again in *Leviathan*, wrote how 'crafty ambitious persons abuse the simple people'. Perhaps it's time to take the late-twentieth-century righting of the wrong done to witchcraft's practitioners over the past millennium and apply it to craft. Isn't someone who is crafty also someone who simply has a way of doing things that is different from our own? Like the witch, the crafty so-and-so is the outsider, the non-conformist, the maverick, the renegade. Their craftiness is about bringing together all their powers to get on in the world outside of the Establishment, or perhaps even despite the Establishment. If we don't already, should we not admire craftiness a little more?

My first brush with an alternative reading of the word crafty was some years ago on a building site in south-east London. I was working with a team of labourers one of whom, Billy, had managed to curry favour with management to the extent that he'd been moved off the labour-intensive job of hand-digging footings and on to what the rest of the team termed 'a cushy little number', checking gravel-laden heavy goods vehicles in and out of the site. Over tea break the next day the others called him a 'crafty son of a bitch'. But it was the way they said it that got me thinking. Yes, they disliked him because of his obsequious behaviour, but there was an element of envy in their tone brought about from an underlying respect for what Billy had achieved. And what Billy had done wasn't morally wrong; it wasn't deceitful or treacherous, he was just trying to get on in life.

I for one would like to claw back the word from its current association with slyness and cunning. For me, this use represents a borderline insult that has its origins in the rise of formal knowledge and an emergent snobbery towards manual artisan skill. It is writers and not makers who create the texts on which dictionary definitions are founded, and this pejorative sense is the result of the tensions that arose between *Homo sapiens* (Man the Wise) and *Homo faber* (Man the Maker) in late-Victorian society. The schism is most beautifully rendered in Thomas Hardy's *Jude the Obscure*, where the protagonist, a highly skilled stonemason, despite being 'a handy man at his trade, an all-round man, as artizans in country-towns are apt to be', is desperate to have his intellectual exertions recognised. Of Christminster (Hardy's fictional Oxford), Jude would say, 'I love the place – although I know how it hates all men like me – the so-called self-taught – how it scorns our laboured acquisitions.'

My aim, then, is to repossess the word crafty from its present-day detractors and reinvest it with some of the qualities it had over a thousand years ago, to embrace it as a form of knowledge, not just a knowledge of making but a knowledge of being. Alfred the Great may well have been a zealous Christian, but he had a darker side, and it shows in his rendering of the legendary heroic craftsman Weland the Smithy, a character famed for his cunning and his *crœft*. Of this mythical metalworker Alfred writes, 'Where now are the bones of the famous and wise goldsmith, Weland? I call him wise, for the man of skill can never lose his cunning, and can no more be deprived of it than the sun may be moved from his station.' As pious as he may have presented himself to an evermore moralising Christian faith, Alfred clearly had an appreciation of the old ways.

Being *crœfty* was about more than just being good with one's hands. In the strategies of translation that Alfred adopts for Boethius's *Consolations of Philosophy*, he unites the concepts of learning and virtue with making by using *crœft* in his translation for the Latin of all three. For Alfred, the labour and work associated with making and doing was comparable to the spiritual strivings of philosophy. It seems we are finally coming back to this notion that making has a spiritual element to it, that making fits within a wider understanding of who we are and where we are going. My definition of *cræft* and *cræfty*, I hope, brings us closer to this.

AGAINST A RISING tide of automation and increasing digital complexity, we are becoming further divorced from the very thing that defines us: we are makers, crafters of things. When our lives once comprised an almost unbroken chain of movements and actions as we interacted physically with the material requirements of our existence, today we stare at screens and we press buttons. When we made things, we accumulated a certain kind of knowledge, we had an awareness and an understanding of how materials worked and how the human form has evolved to create from them. With the severance from this ability we're in danger of losing touch with a knowledge base that allows us to convert raw materials into useful objects, a hand-eye-head-heart-body co-ordination that furnishes us with a meaningful understanding of the materiality of our world. Some people call this knowledge know-how to distinguish it from formal knowledge, the knowledge of principles. But you could call it *cræft*. It is a wisdom that furnishes the practitioner with a certain power.

We appear to have created a society that looks disparagingly on people who use their hands to earn a living. Nowhere is this more pronounced than in the educational system where value is placed on the learning of principles rather than on learning through doing. The implication is that people who work manually on a day-to-day basis don't have the intelligence to sit at a computer. To be fair, our machine-driven world of manufacture has meant that a plethora of processes that were once skilfully undertaken by hand are now conducted with the flick of a switch, the pressing of a button or the easing of a lever.

So we can't argue that people who electronically use machines to make things are any better than computer operatives. But as a wider consequence, fabrication, construction, energy, waste and by-product are largely monetary abstractions to a society of non-makers. It occurred to me that if we spent more time individually converting raw materials into useful objects, we might be better placed to contextualise the challenges that face a society addicted to excessive and often conspicuous consumption. Perhaps more importantly, we might be a little bit happier.

In many ways this book is an attempt to pare back to the basics of human existence and explore worlds in which the sustenance demands of life were met by the endeavours of our own hands. I realise that a linear narrative of machine manufacture replacing handcrafting is fraught with contradictions, and that many of the crafts discussed in this book have arguably held a more illustrious position in society *after* the major scientific developments of the last two centuries. I'm not against machines in principle, and I certainly don't see craft as a simple dichotomy of man versus machine. But the use of machines for manufacture can create a social and economic jarring, the results of which are inundation, devaluation, waste and inequality.

I've always thought that creating machines for tasks we could just as easily accomplish ourselves is unforgiveable – the battery-powered pepper mill comes in for particularly venomous scorn in our household. It's a potent symbol of a society that's going backwards. And to those who argue verbatim the advertising spiel that a battery-powered toothbrush reaches parts that a regular tooth brush can't, I would say: leave your critical faculties in a glass jar overnight, like a set of false teeth. But while some machines are clever, the net result of our using them is that we become lazy, stupid, desensitised and disengaged. We must never lose sight of the fact that the most intelligently designed, the most versatile and the most complex piece of kit we have at our disposal is our own body. As John Ruskin put it in 1859, in our hands, we have 'the subtlest of all machines'.

I AM AN ARCHAEOLOGIST. Archaeology is the study of anything that has been made by the human hand and, as such, the principles of the discipline can be extended to a whole range of phenomena from tools and objects to buildings, monuments and landscapes. By examining their physical form and spatial contexts we create narratives about past societies and seek to understand them through their creators' own hands. This approach can also be extended to modern materials and to an archaeology of modern Britain; we can examine our material culture and draw conclusions about who we are and what we value. I worked in the construction industry as a commercial (as opposed to research) archaeologist for over five years. Nowadays, for any building development, no matter how big or small, a requirement to undertake archaeological work is written into the planning process; if there is a chance that archaeological data will be destroyed, through the installation of foundations and service trenches, the company is required to initiate archaeological work.

It was in this environment that I put my university-learned theory into hard commercial practice as I set about excavating and recording archaeological finds from a rich variety of periods and in a number of fascinating locations. These were my salad days. I revelled in becoming what's known in the trade as a circuit digger, a hired trowel. I travelled wherever the work took me across the south-east of England. I dossed down in old caravans, church towers, derelict beach huts, tents and rundown B & Bs. I lived out of the back of my car, I dug hard and I drank harder. I was free.

Eventually, the lifestyle got to me. I was beginning to get a bit disillusioned with developer-led archaeology. Undercutting was rife, costs and wages were relentlessly driven down, and as a consequence the execution of the archaeology was often crude and expedient. There is also a certain myopia that comes with having one's head stuck in a hole all week. I pulled some fascinating archaeological finds out of the ground, but in many ways I was peering at the past through a tiny conceptual window and seeing only fragments of the possessions that had once belonged to the people who had lived there. I began to struggle with some of the interpretative models that other archaeologists were coming up with because they seemed to me to require such a monumental interpretative leap in order to get from dust and bones to actual process. We were really good at describing *what* happened in the past, but the *how* and *why* either evaded us or was summed up in a series of trite generalisations and overconfident conclusions.

As part of this self-reflection, I didn't set about applying complex critical theories to archaeological data in order to better understand the link between pattern and process. Instead, I applied for a job on a historical farm that was to be the setting for a BBC documentary series about life in rural Britain in the year 1620. The idea was simple: you take a re-enactor and two archaeologists and place them on a farm and entrust them to run it as it would have been back in the day. I'd be lying if I said there wasn't some attraction at the prospect of being on TV, and as my old car rattled down the M4 in the direction of the seventeenth century, with Buck Owens' classic 'Act Naturally' blaring out on the stereo, I considered what fame and fortune awaited me. But deep down, I was in search of a 'dwelling perspective' on the past, a more complete understanding of the link between the wider environment and the archaeological record as a function of human inhabitation and interaction. I wanted to become the avatar within my own theoretical world but, most importantly, I also wanted to see how the archaeological record was created in the first place.

When *Tales from the Green Valley* aired in 2005 it proved something of a surprise success story. During production, I'd often wondered to myself who on earth would want to watch a bunch of cranky, oddball re-enactors and archaeologists bimbling around in costume, pretending to live in the past. But I didn't care too much because I was spending nearly every single hour of every day immersed in historical farming. I was tending, ploughing, scything, chopping, sweeping, hedging, sowing, walling, slicing, chiselling, digging, sharpening, thatching, shovelling; the list is almost endless. Most significantly, I was watching with an archaeological eye how my actions were altering and reconfiguring the material environment around me. For a period my interests and passions overlapped with the commissioners at BBC Two, and over the course of the next six years I made a substantial contribution to the making of *Victorian Farm, Edwardian Farm* and *Wartime Farm.* Though shot in a slightly different style, these series were essentially based on the same premise as *Tales from the Green Valley.* I was fortunate enough to try my hand at a huge range of crafts and, in forcing myself to implement them on the farms, I gained insights I would never have found if I'd been doing them for fun in my own back garden. When I made a traditional hay rake, I did it not to hang decoratively on my shed wall but to use in earnest, gathering in my own crop of hand-mown hay.

At about the same time as the TV work took off, I decided that I wanted to return to study another of my passions – historic and archaeological landscapes. I won a scholarship to undertake a doctorate at the University of Winchester, and in the downtime between productions I spent an enormous amount of time wandering the downlands of southern England exploring the ancient and medieval landscape of Wessex. For a period of nearly ten years, on one hand I was immersed in crafts, and on the other, landscapes. And I began to understand the reciprocity between them. Crafts, through their need for raw materials, created patterns in the landscape, while landscapes determined the nature and character of the craft life. In a neat circle, crafted objects also helped shape the landscape. The surrounding environment could be read as a record of the lives and, critically, the work of people in the past. In my work for the various BBC farm productions, I had become one of the characters in Bruegel the Elder's *The Harvesters* (1565), a painting that brilliantly illustrates the degree to which people were part of a complex interaction with plants, animals and the built environment, all set out in the tableau of a *cræfted* landscape.

IT WOULD BE impossible to contemplate a book on crafts without at some point engaging with the Arts and Crafts movement of the late nineteenth century. Popularly heralded as a reaction to the industrialised mass manufacture and often vulgar consumerism of mid-Victorian Britain, it was influenced in its design ethos by the medieval revivalism of an earlier generation of architects, such as Augustus Pugin and George Gilbert Scott. By the 1880s, the time was very definitely right for the emergence of a new brand of manufacturing. Levels of disposable incomes among the rising middle class were the highest they'd ever been, and a growing support for more socialist ways of thinking created the political space to conceive new modes of production.

The first Arts and Crafts exhibition, held in 1888 at the New Gallery in London's Regent Street, did much to lift craft into a state of self-perception, reflecting an intellectual engagement with how making fits into society, culture and the economy. It found its place during a period of regional growth in craft awareness – not so much a revival but an early attempt at conservation. For the first time since the medieval period, guilds and societies were founded in order to bind together designers, artists, architects and craftsmen.

One of their chief ambitions was to address what many in the movement felt was a need to reform the design process. But at the same time there was a very real desire to return the maker – the craftsman – to the process of fabrication. John Ruskin, one of the movement's founding members, was the first to place an emphasis on the true value in an object or building being derived from the pleasure taken in creating it; a key tenet of the Arts and Crafts movement was the notion of harmony between designer and craftsman, producing attractive, well-made, affordable objects for everyday use.

The movement encompassed a number of different practitioners with a wide range of ideals and beliefs but, in general, they followed shared interests in the use of local materials, vernacular styles, a nod to the past (medievalism), and simplicity and honesty in design. Perhaps the most influential of all its exponents was William Morris, who was born into a comfortably well-off family in Walthamstow in 1834. A writer, lecturer and educationalist who went on to become a designer, craftsman and poet, by the 1850s he was ill at ease with what he considered to be a society in a state of decay and disorder. Like Ruskin before him, he saw the answers in the harmonising of design and crafting in the production of day-to-day goods from natural raw materials. Morris drew around him an eclectic mix of artists and designers, including the painters Dante Gabriel Rossetti and Edward Burne-Jones and the architect Philip Webb. Their collaborative venture, Red House in Upton, Kent, a residence commissioned by Morris, was to seed a business enterprise that fused decorative arts with fine arts and architecture, all delivered through a handcrafted sensibility.

The Arts and Crafts movement had always harboured a

commercial ambition for their designs and products. At first, this took the tone of moralising lectures and pamphlets on the case for good design and handcrafted objects, but over time they found themselves sleeping with the enemy. To compete with mass manufacture, design companies reliant on handcrafting simply couldn't match factory production for volume and still provide affordability, so machines were increasingly employed in certain manufacturing processes. In return, industrial producers were attracted to the marketability of the unique selling opportunities the Arts and Crafts style offered. There were other contradictions too. Objects laboriously produced from start to finish using only human hands were too expensive for anyone but the wealthiest. Worse still, the wealthiest in society were invariably the industrialists whose money was made in the very factory conditions the Arts and Crafts luminaries deplored.

In a cycle of contradictory irony, the captains of industry used Arts and Crafts objects to overtly display their wealth and status, while the captains of Arts and Crafts relied on industrial money for their patronage. In light of the movement's profound cultural impact it seems specious to do it down. Ultimately, its leaders had a radical effect on design principles coming into the twentieth century. They vastly broadened the ranges of techniques used in the making of everyday domestic items, they gave a much needed aesthetic boost to vulgar late-Victorian tastes, and placed a sense of their national past more centrally in how a building or object should be conceived. But, for me, one of the most intriguing episodes in the Arts and Crafts story is that of the creation and relocation of the School and Guild of Handicraft by Charles Robert Ashbee.

Ashbee was much more a designer and businessman than he was a craftsman. He schooled himself in Ruskin's doctrine and, influenced by Morris, had political leanings towards socialist and collectivist ideals. A designer in residence at Toynbee Hall in East London, Ashbee was strongly governed by a desire to see his created objects set within a framework of self-sufficiency and an integration with nature. For this aspect of his enterprise, the countryside was the fitting – and only – place where this could be achieved. As a consequence, the workshop set up in 1888 in Commercial Road in London's East End, was moved to Essex House in Ilford in 1891, and was finally relocated to the rural backwater of Chipping Campden, Gloucestershire, in 1902. Here, among a relatively warm reception from the local community, Ashbee's Guild of Handicraft had limited success.

His vision, in retreating to this countryside idyll, was to improve the standard of craftsmanship as well as the status of the craftsman, but as it turned out the best place to establish a bucolic ideal was not necessarily the best place from which to sell high-end designer metalwork, jewellery, enamels and wrought copper and iron furniture. Ashbee's Gloucestershire concern lacked the centrally placed retail outlet the Guild had enjoyed in Brook Street, Mayfair, where he could attract passing wealthy patrons. By 1907, the business was all but over, yet the dream that objects could be enjoyably produced in an environment of communal living, and the profits from sales shared evenly, very much lived on.

I admire Ashbee and his enterprise, though he may have been before his time. Today, with internet access to global markets and a delivery network to match, I have no doubt that his communal workshops would have been a standout success. Critically, what I like about him was his desire to place craft within a wider social and economic setting – even if his chief fault was that he did this too literally.

BUT WHAT ABOUT the skill of making? It's easy to stand back and marvel at a craftsman masterfully manipulating tools and materials, but what is that special something a particular craftsman has that results in such beautiful objects? Think too much about sheer talent and it can quickly escalate into the realms of the mysterious and the magical. But what is this ineffable ability, and is it even definable? Tacit knowledge plays a substantial role in the way we teach, learn and practise in a wide variety of professions. It is this indefinable knowledge that George Sturt in his classic 1923 ethnographic study of a particular group of craftsmen, *The Wheelwright's Shop*, repeatedly referred to as 'real knowledge', while at the same time finding it almost impossible to define with any precision. I also struggle to convey in words the actual crafting of an object. Seeing is believing, and words alone are not enough to truly express what it is to create skilfully.

David Pye, whose *Nature and Art of Workmanship* represents one of the most authoritative commentaries on skill, was of the opinion that workmanship (his term for 'skill') is at least susceptible to rational examination, that it can be broken down into a series of movements and conscious processes. I'm reminded of the work of Michael Brian Schiffer, a professor of anthropology and an eminent behavioural archaeologist, whose contributions to archaeology I was at pains to understand as an unversed student of archaeological theory in the mid-1990s. In academic writing so dry I could almost feel my eyes desiccating on the page, Schiffer appeared to demonstrate that the conditions for success in the production of flint tools could be modelled through the scientific analysis of replication experiments. In essence, what worked and what didn't could be modelled, and the cognitive processes behind these decisions could be inferred. I have no doubt that similar approaches could be adopted in the studies of modern craft processes, and that we could begin to characterise in meticulous scientific detail the 'real knowledge' with which Sturt, among others, was so fascinated and perplexed: how craftsmen arrive at the best possible method for exacting the perfect object.

We need only look at the world of sport to see how almost every aspect, from the mental and tactical through to the nutritional and physiological, is placed under staggering levels of scientific scrutiny in a bid to gain advantage. But even if we could successfully describe and map the decisions craftspeople make in the processes of creating, can we really get to the value judgements and motivational desires that lie behind them? More importantly, would we really understand them any better than if we practised them ourselves? There is undoubtedly a healthy compromise somewhere between the blind admiration of the untrained onlooker and the over-analysis of the cognitive scientist.

AT THE HEIGHT of the Industrial Revolution, as machine power spread its tentacles through all aspects of production, Britain and its colonies were flooded with a veritable cornucopia of consumables. The Great Exhibition of 1851, held in the giant glasshouse at London's Crystal Palace, was intended as a celebration of this industrial prowess, but in certain quarters there was unease at the emerging culture of mass consumption and, in particular, the effect the increasing use of machinery was having on the skill levels of British workers. Few critics were more vehement in their attacks on industrialism than John Ruskin, who voiced particular concern over the working prospects of the craftsman in the face of ever more mechanisation. Ruskin talked of the 'degradation of the operative into a machine, which, more than any other evil of the times, is leading the mass of nations everywhere into vain, incoherent, destructive struggling for freedom'. Strong words indeed. Of the labourer, Ruskin implored his contemporaries to see that he was not 'activated by steam, magnetism, gravitation, or any other agent of calculable force', and that his real motivating power was his 'soul'. Machines were therefore not just perceived as a danger to the livelihoods of the craft community, they threatened to undermine the very fabric of British life.

In retrospect, the interrelationship between man, machine and manufacture was far more complicated than Ruskin initially conceived it. First, many machines were saving workers from some of the more laborious and unsavoury aspects of industrial production and were welcomed by folk working on the factory floor. Second, in some instances, machines were undertaking new work in manufacturing contexts – that is, work that wasn't previously carried out by human hands. Third, even in those times of increasing mechanisation, there was still a requirement for skilled manual labour to work the machines. As manufacture expanded in the late nineteenth century there were, in fact, a greater number of opportunities for crafts to develop. It has been argued that the juxtaposition between traditional forms of manufacture and the emerging industrial complexes is what created our modern notion of craft in the first place – it was only when machines came along that the distinction needed to be made.

There is, then, no tidy historical narrative that allows us to make a clear distinction, in mechanical terms, of when craft stops and machine manufacture begins. In which case, it might be more useful to consider the point at which a tool becomes a machine. John Harris's *Lexicon Technicum, Or, A Universal English Dictionary of Arts and Sciences,* published in 1704, provides one of the earliest definitions of a machine as 'the Lever, the Balance, the Wedge, or inclined Plane, Screw and the Pulley'. One could say, therefore, that a pair of scissors represents a form of machine. Two blades, effectively levered against each other, allow the operative to cut in a controlled fashion without the need for a cutting bench or stabilising brace. There is an element of what Howard Risatti in *Theory of Craft* calls 'mechanical advantage' derived out of a 'system formed and connected to alter, transmit, and redirect applied forces'. But in this particular example it isn't in the replacement of manipulative skill, for the scissor operator still has to use a series of bodily controlled movements to ensure that the cut is made along the desired line.

Two succinct examples give some sense of where I see craft positioned in relation to these bodily movements – the use of tools, machines, power and the overriding context of the work. I'll start with topiary, the craft of pruning shrubs into decorative shapes. In any productive flower, vegetable or fruit garden, a tightly clipped hedge is almost a necessity, and the more hedged borders one can afford to maintain the better. While their roots can stabilise and help to contain garden soil, a tight-knit hedge's chief benefit lies in its role as a screen of dense foliage. This can provide a barrier against wind-borne weed seed ingress. It can also offer a wind shield to more delicate plants in the garden and, in an age before asphalt road surfaces, in the drier times of the year the hedge would stop the worst of the dust being picked up and blown into the garden. It was not unknown in the Tudor period for these well-manicured hedges and bushes to be used for the drying of laundry. And, as any amateur ornithologist will tell you, a good hedge attracts garden birds, which in turn do an excellent job of keeping insects at bay. Clip the wings of geese, ducks or chickens and a tight hedge of waist height will prevent them escaping from a contained daytime run. So, the art of topiary is really just an extension of a fundamental garden craft.

Pliny the Younger, writing in the first century AD, informs us that the gardens of Tuscany were adorned with the representations of different animals shaped from box hedge. The tools used by these early topiarists are likely to have been the sickle – or more

34

specifically a hedging hook – and the sprung shears. The sickle was swung in a slicing motion, a technique sometimes known as brushing, to swipe out the larger shoots, while the fine pruning of the foliage would be conducted with the shears. Sprung shears are forged from the same length of metal, blades are hammered out at each end and then the metal is bent round on itself so that the blades oppose each other in a sprung-like fashion.

The form of opposing two blades through a pivoting pin was invented in the medieval period, and by 1760, when the manufacturer William Whiteley & Sons was founded in Sheffield, scissors were being sold in substantial numbers in Britain. It wasn't until well into the nineteenth century that garden shears based on this principle were more widely available. There is a short leap, in technical terms, from these to the finger-bar shears that require much the same kinaesthetic sensibility, what Risatti defined as the 'sensation of bodily presence or movement'. There is a point at which the process becomes mechanised: mechanical advantage is gained through the gearing up of power by means of a hand crank, operated in a circular motion, which powers the oscillating finger-bar blades against each other. In its final form this mechanism is powered by an external force and we arrive at the hedge trimmers that can be heard chattering away on warm spring weekends in the suburbs of the developed world.

The craft of trimming hedges can be broken down into three physical functions. Number one is the application of power. Number two is the kinaesthetic sensibility that enables us to shape our body, arms and hands into a position that allows us to achieve number three, the act of cutting. In the first three examples in the evolution of hedge-trimming equipment, physical functions one, two and three are all achieved through the human body. In the fourth example, the function of cutting has largely been reduced to a mechanical action, a redirection of the transmission of power. But it's still the operative, via the action of repetitive hand-cranking, who powers the machine.

The illustration of the fifth and final phase in the evolution is missing some details – the power lead, plug, socket, domestic electrical circuit, National Grid and power station required to make it actually work. In this final example, the machine is undertaking the act of cutting, electricity is providing the power and the operative is reduced to guiding the machine in the direction and manner they choose. As such, a hedge can still be crafted with this implement. But is it being *cræft*ed? Does this demonstrate the knowledge, power and resourcefulness of *cræft*? At what point are the complexity of the engineering and the embedded carbon cost of the machine and how it's powered offset by the advantage gained from using it over the four other examples in this illustration? And at what point does the social and economic context have to change to tip the scales in favour of a return, if not to sprung shears but to examples two, three and four? This is about resilience and sustainability as much as it's about setting the benchmark for when crafting begins and ends. Perhaps harshly, I would not consider a topiarist who uses electric hedge trimmers a true craftsman on the simple grounds that the tool mutes their level of engagement with the material properties of the entity they are working.

Not all of us have hedges in regular need of tending, so let's look at an example closer to home. This is the craft of getting from A to B. Risatti uses the example of the bicycle to illustrate that through a series of mechanisms we can gain mechanical advantage – both in the redirection of power generated through the downward act of pedalling and through subsequent gearing systems. But there is a cost to this particular machine. Bikes require a certain condition of surface to operate effectively on, such as tarmac. They also have to be manufactured and maintained at a cost (tyres, brake pads and oil for moving parts). While the bicycle might be considered an example of a machine that has improved our quality of life immeasurably, it also removes us from a natural state. The greater velocity allowed by the mechanical advantage places us in a more exposed and vulnerable position. If for whatever reason the rider were to part company with the bike in motion, the body is not designed to impact on hard surfaces at these increased speeds, and the consequences can be severe, if not fatal.

It might seem specious to criticise the bicycle. After all, unlike the motor car, it uses human power to propel it. Without a bike I could never have done my paper round as a kid. In which case, I wouldn't have earned pocket money, the paper shop and newspaper magnate would have sold fewer papers, and our customers would have been less abreast of current affairs. In short, everybody would have lost out. The point I'm trying to make here is that in the act of cycling there is a level of disengagement with the physical reality of getting from A to B. We may save time, and, in my case, earn some precious needed cash as a teenager. But will it always equate to the cost of increasing physical jeopardy, the capital cost of bicycle manufacture and maintenance, and the manner in which we are disengaging with the material world around us? In this context, to walk might be seen as being more *cræfty*.

In these two examples I've tried to create a link between an action or craft and its wider socio-economic context, its landscape of use, and to judge it on those terms for its efficacy, fittingness, lasting value and, for some, its beauty. It's about more than just making. The goal, in being *cræfty*, is not to use as much as possible of the technology and resources you have at your disposal but to use as little as possible in relation to the job that needs undertaking. This is the resourcefulness in *cræft*. Having physical adeptness, strength and fitness represents the power in *cræft*. And finally, understanding the materials, making critical decisions about how to approach the work, and factoring in wider financial and time constraints represents the knowledge in *cræft*.

2

MAKING HAY

IF MAKING DEFINES us as human beings, then I'll begin with the making of one of the most basic products of human endeavour. Today, western society is almost as removed from haymaking as we are from the fifth millennium BC. But if we were ever to successfully domesticate bovine or ovine species and exploit them for their meat, milk and skins, there would always have been a need to provide feed throughout the year. No part of the world can avoid the seasons, and the grasses, plants and herbs that are abundant as a food supply in one half of the year can be exhausted very quickly by ruminant animals as the supply goes to seed, withers and lies dormant for the second half of the year. Migration was the means by which grazing animals negotiated this, travelling vast distances from exhausted pastures to fields of plenty in another part of a continent. As hunter-gatherers, humans undoubtedly travelled with them in a parallel migration. But as we began to establish a more sedentary existence, we had to make up for the shortfall in food during the winter (or, in some cases, summer) months. For agrarian communities the world over, making hay was the answer.

Hay and its making provides us with a shared heritage. In the forging of it in the modern mind as a timeless and iconic act of our rural past, we've probably been a little too guilty of undermining its centrality to the human story. I don't think it's too much of an exaggeration to say that whole empires and civilisations have been built on hay. Without it, we would have struggled to sustain large herd and flock numbers. Economies based on hide, wool, skins, bone, meat and milk would have suffered. The fields where crops were grown would not have yielded the same levels: livestock integrated into the farming system provided the precious manure to support more intensive systems of crop husbandry. It is this intensity that produces the surplus. And so capital, wealth and power stem from it.

So what exactly is hay? Essentially, it's cut grass. Animals feed off grass, so if you can find a way of taking that grass and storing it in large volumes you can share it out over the winter months for your cattle or sheep until such time as the temperature changes and new shoots of grass emerge the following spring. Simple. But not so simple. Because if you cut the grass and bundle it up for store it would rot, as any organic matter would, and do more harm than good if fed to hungry animals. For grass to be used as winter fodder it needs to be made into hay. More to the point, if you are to become a successful farmer, skill, resourcefulness and knowledge need to come into play. There is undoubtedly a *cræft* to haymaking. And the tools for the job have always been of the simplest variety, consisting of something to cut the grass – a scythe – and then something to handle it as it's made into hay – a pitchfork.

We are all familiar with the shape and form of the iron or steel-headed pitchfork. Certainly, by the twentieth century, these had become widespread across Europe and America as a multi-purpose tool around the farm. But its adoption is tied into forms of mechanical harvesting where the volume increase – and particularly the compaction of both hay and straw into bales – required considerable strength in the tines of the fork used to lift them. For the making of hay by hand, however, the iron-headed pitchfork is a rather heavy and clumsy beast. In fact, it's borderline dangerous. We always used to joke on the farm that the pitchfork, with its sharp tines swung liberally around at a distance of some four or five feet from the body, was only ever an accident waiting to happen. For the techniques of tossing, tedding and pitching loose hay, a wooden-tined fork is the tool of preference – because of its lightness and the safety needs of working around animals and other humans. I've worked with wooden forks of a composite nature where the tines are fastened to a cross bar fixed to a long handle. But by far the most superior wooden pitchforks are made from a single piece of wood.

So effective are these forks that there is a factory in France that still makes them. When I visited the Cévennes region of southern France about ten years ago, I encountered many people travelling the dusty tracks and highways in the service of God, en route to the shrine of St James at Santiago de Compostela in Galicia, north-west Spain. My pilgrimage to Le Conservatoire de la Fourche (the Conservatory of the Fork) in Sauve was of an altogether different variety. This was a place that has been making wooden pitchforks since the twelfth century, and I had come to pay homage to a tradition of fork-making that is at least eight hundred years old.

The Cévennes region, and particularly the area around Sauve, traces its origins as the spiritual home of wooden fork-making back to an act of classic French self-preservationism when, in the seventeenth century, the leader of the fellowship of fork-makers managed to persuade the king to grant a monopoly to the people of the region. France as a whole probably benefited, both because the fork-makers of the Sauve were so accomplished but as much because the area produces such fantastic raw materials for the purpose. As with any true craft product, the environmental character of the region is crucial: the forks are made from the nettle tree, which is relatively fast-growing and has a smooth grey bark and sharp-toothed narrow leaves. They thrive on thin, loamy, well-drained and nutritionally poor soils, and for this the steep declivities and limestone shelves of the gorges and river basins of the Cévennes provide the perfect habitat.

Although the factory, which doubles as a museum, was incredibly interesting, it was by the groves of young nettle trees, the arboreal equivalent of a finishing school, that I was most enchanted. It's here that the real work is done, where each sapling

43

is trained into shape, ready for its vocation as a trusty pitchfork. Like a fine whisky, the production of a wooden pitchfork takes at least ten years. It begins with the planting out of baby saplings reared from seed. These are trimmed near their base at about five years of age. The following spring a number of shoots emerge from the stump where the sapling has been severed. Over the course of the next five or so years, the fork-maker trims and manipulates these shoots in such a way as to create the prongs of the future fork. Once of age, the young trees are cut down, cooked in an oven, bark stripped and, using more heat, further manipulated into the correct shape. With no fixings, screws, artificial joins or glue, this implement is incredibly strong, lightweight and made from a wood that, when seasoned, is famously resistant to rot.

In some ways the finished article beautifully reflects the Cévennes region, an area particularly celebrated for the continuous negotiation between man and nature. With the usual rebellious spirit that comes with the long-held occupancy of an austere landscape – a political resilience born out of an ecological resilience – the people of Sauve and its hinterland are proud of their ingenuity. Their wooden forks serve as a timely reminder of our close links to landscape, and our dependency on the immediate world around us, not just on what that landscape has to offer in terms of resources and raw materials, but on the crafts skills we have designed over the years to sustain ourselves.

THE SCYTHE IS now garnering fans the world over. While it may never reach the same elevated status it had in the medieval period, it's experiencing a renaissance in the west, with many people undergoing the same epiphanic journey as I did on that late April morning. In the US not only has it become an icon of the backlash against the all-American manicured lawn, it is also the must-have tool for smallholders and environmentalists. There is no doubt in my mind that this rebirth of the scythe's popularity is ultimately underpinned by a deep philosophical stance against our increasing reliance on fossil fuels.

Central to this philosophy is David Tresemer's *The Scythe Book* (1981) – the bible for hand-haymakers. A generation before I first picked up a scythe, Tresemer had made the evolutionary step from over-engineered, petrol-reliant machinery to a simple and timeless substitute, and in so doing eschewed his mechanical mowers in favour of the humble scythe. I have come to *The Scythe Book* late in life and have found it comforting that my own journey has not been an isolated and irrational throwback to a lost world. Like me, Tresemer believed that 'a scythe can perform a moderately sized task in the same amount of time it would take to fetch, attach, adapt, and repair a mechanical substitute', and that, 'Maintenance of the machine means money spent; maintenance of the human body means health gained.' But he is also a realist (where I am sometimes not) and stresses that, 'romance aside', the scythe 'must make such an effective use of a person's time and energy that it is competitive with other means for accomplishing the task at hand'. It is these sentiments that have seen this classic book rise to the status of a sacred text among the growing army of scythe-wielders.

And like me, Tresemer made the transition from the heavy American (or in my case, English) scythe to the lighter and more forgiving European scythe. But it's not entirely clear in the book exactly when he turned to full-blown haymaking. As anyone who has ever taken pleasure in cutting grass with a scythe will tell you, it's a logical progression to find yourself raking at the freshly cut grass and pondering the challenge of how you would then go about turning these cuttings into hay. It was the year after my early scything endeavours that I first embarked on a haymaking project, and although I didn't have any livestock to feed, I was intent on conducting my own little experiment to explore just what it meant to make hay by hand. Looking back, I clearly had more time at my disposal than sense. I had returned home from an extended archaeological excavation abroad to find the garden under siege, not only by the usual hogweed, cow parsley and dock leaf but by the lawn itself, which in my absence had developed into a small meadow of mixed grasses and flowers. It was beautiful to look at, swaying in the late June breeze and danced upon by all manner of butterflies and bees. But to my new haymaking eyes it represented a crop to be harvested.

For the next three days I was gifted the perfect weather: hot and dry with a gentle breeze. I cut systematically, laying the swathe to the ground, and with every swing of the scythe I became more and more entranced by the journey back in time I was embarking on. Once cut, I spent the next two days hovering around the hay, turning it gently with an old garden fork, raking it into rows of an evening and, in an act of mock knowingness, testing its sweetness between my teeth. At the end of day three I had decided it was 'made'. I gathered it up into the loft space of one of my outbuildings and cracked open the cider. I was mellowed, philosophical and relaxed. But more than anything I felt connected – with a place, with a past and with myself. My passion for haymaking was born.

IT WASN'T UNTIL much later in my small-scale farming career that I came to consider haymaking as more than just an agricultural practice with a deep time signature. My revelation that it was, in fact, a craft – and one of the original *cræft*s – was triggered by a particular entity I encountered while thrashing an old Land Rover up the A303 in the direction of London. It was an evening in late June, during a summer that seemed to promise glorious sunshine but never really delivered. They say that you make hay when the sun shines, and any farmer keen to bring in a hay crop that year had to make do with the odd short spell of blustering wind and bright sunshine in predominantly showery conditions. That day a strong south-westerly wind coaxed the Landie up the A303, and the crimson red horizon in my rear-view mirror did little to dissipate my frustration: there was clear sky over the Atlantic and it was coming this way. My own haymaking ambitions were also hanging in the balance for, with a busted mower and the potential for a scorcher on the morrow, the last thing I needed was a day away from the hayfields.

As I rattled past Ilminster I entered a stretch of arterial bypass, a ring road that hung tight like a poacher's snare around the neck, throttling the wealth out of this once bustling market town. Where the residual ground rose up on either side of this cutting, my eyes were caught by something on top of the expansive embankments. Perched on the crest of these sidings, regularly spaced and seemingly of deliberate construction, were what I can only describe as piles of cut grass. As my fellow A303ers hurtled by oblivious, I slowed down, leaning forward on the steering wheel and peering at them through the wiped arcs of an otherwise filthy windscreen. I cursed myself for not having my camera to hand, sped up and continued on towards London. So why, you might ask, had these stacks of hay piqued my interest? The answer is that these innocuous piles of grass looked for all the world like what the haymakers of old called 'haycocks'.

In the years running up to this summer, I'd become obsessed with what was once romantically termed the art of haymaking. In a bid to understand the history and archaeology of haymaking, I travelled the west coast of Britain from Scotland to Cornwall. I camped out in the mountains of Asturias in northern Spain and trekked the fjords of Iceland. The basic premise was that the further west you go in the British Isles, the craftier you need to be to successfully bring in a crop of hay. To be more precise, the closer you get to the inclement weather of the Atlantic seaboard, the more difficult it becomes to find the windows of hot and dry weather during which to bake your cut grass and turn it into hay.

Whereas in the east of England endless roasting summer days afforded idyllic scenes of plenty and haymaking was a calm, relaxed and almost pedestrian ritual, similar bounties in the rainy west would require all the guile, tactical nous, quick thinking and creativity of a master. The rewards were there to be had, though, for in the west the wetter climate and warmer air brought in on the jet stream could produce early and especially luscious grass. Get a cut in early in the spring and you might double your money with a bonus cut later in the summer – weather permitting. Haymaking in these conditions involves wrestling with the Atlantic weather, and learning to play poker with nature's titanic forces. It isn't simply a case of cutting the grass, hoping the hot sunshine turns up to dry it, collecting it into bales and returning smugly to your farmhouse for tea and cake. It's a much more complicated process and one where crops – and farmers – could be ruined by a lack of *cræft*, poor decisions and bad timing.

There are so many variables in the craft of haymaking that trying to explain them all here would be like attempting to compress a manual on Test cricket into a few paragraphs. However, there are a few key principles. Cut too early in the year, for instance, and you run the risk of bringing in an immature crop: thin and innutritious to the animals eating it throughout the winter months. Cut too late and the grass will potentially have gone to seed; the precious nutrients the haymaker needs to capture in the stems of the plants will have migrated to the seed head. In the process of scything and mowing late into the summer, the dry brittleness of ageing plants can scatter their seeds about the meadow floor, providing a feast for wild birds but of little use to the farm. The effort of bringing in the resulting fodder – grass denuded of its nutrients and seeds – is scarcely worth the calorific value of the final product. Moreover, if you let a hay crop get too thick and heavy, a squall could flatten it and make it impossible to mow, clogging the cutter or breaking the scyther's back.

One of the greatest variables in haymaking by hand – aside from the capricious weather of the British summer – is the labour you have at your disposal. The number of hands you have to help can dramatically impact on the manner in which you make your hay. For instance, it can be the deciding factor on whether you choose to leave the cut hay in the swathe or select instead to 'break it out'. To the urbanite uninitiated in the arts of historical haymaking, I often make a rather crude comparison between this decision and the cooking of a sausage. The swathe is basically the row of cut hay left as it falls from the cutting process. This is your sausage. If you choose to let the swathe bake off in the hot sun, the outer surface will cook off quicker than the inside and the underside. The trick, as with cooking a sausage, is to turn it just at the point that you have cooked it off to the centre point of the swathe. Then you turn it and start cooking it from the other side. It takes time, and there is always the danger of overcooking, but other than having to wait for it to cook, all you've had to do is turn it once.

But what if you wanted to cook your sausage faster – because you were in a hurry – and you had the required labour to speed up the process? As a student in London I used to frequent a café on Turnpike Lane almost every Sunday morning. At this time of the week the café was at its busiest as various waifs and strays piled in for the traditional British cure for one too many drinks on a Saturday night. I used to watch the kitchen staff frantically knocking out breakfast after breakfast to a waiting crowd of bleary-eyed partygoers. With the exception of the sausage, everything else – bacon, eggs, tomatoes – could be fried fairly speedily. But the chef had found a clever way to speed up the process of cooking the sausage, and this was to slice it clean down the middle and cook it from the inside out as well. This meant that sometimes the sausage lost some of its succulence, but I didn't mind because it meant that I got my breakfast a little bit quicker.

With hay, if you have the labour you can break it out of the swathe and cook it off a little faster. Of course, extending the metaphor, you could completely break out the sausage into its smallest parts – the mince – and spread it as thinly as possible across the frying pan. Hay treated in the same fashion can cook off super quick and the 'making' can be accomplished in as little as a day. But it requires lots of hands on deck – both for the breaking out and for the subsequent raking up and handling – and it exposes the crops to the elements that bit more, making it less resistant to a brief summer shower or a sudden spell of roasting sun.

How much labour you can call on also impacts on how much you choose to cut before you start the making. Mow all the grass available for haymaking and, should the weather turn, you stand to lose the lot before you can get it in. In variable conditions, it's better to take on bite-size chunks to be certain of manageable quantities. You also need to consider issues such as the proximity of the fields and the other jobs that need doing on the farm. As with so many farming practices, modern technology has deprived the farmer of a precious skill base. Today we are far more likely to make silage. The main difference between silage and hay is that hay is dried to preserve it and silage is pickled, by being kept in oxygen-starved conditions so it doesn't decompose. This way it keeps its nutritional value with the added benefit, over hay, of retaining its succulence.

Crucially, the popularity of this method of converting grass into animal feed is largely dependent on the practice of black polythene wrapping in the field. This impermeable by-product of the oil industry provides the anaerobic conditions required for fermentation while at the same time protecting the silage from wet weather. The wrapped polythene bales can sit out in the field until the farmer, at a time of his choosing, sends out an army of ten-tonne trucks and loaders to handle them into bale stacks the size of apartment blocks.

So not only has the need to get the crop in before the weather turns been obviated by making silage but there isn't quite the requirement for longish spells of dry and sunny weather to turn grass into hay. For silage, the grass doesn't need to dry off too much – perhaps a day to burn off any surface moisture and dew – before it's wrapped. The ease of silage making hasn't stopped farmers from making hay, though: it remains a staple feed for an ever growing national stable of riding, racing and jumping horses, which require much less in the way of succulence through the winter months. But it's probably fair to say that the true craft of haymaking is on the wane, for while the same disastrous errors that afflicted the farmer of the past can still impact on the successful making of hay in the field today, heavy machinery affords the luxury of not having to confront just how much of a craft it was in the old days to bring in the hay.

The power and speed of modern tractors, cutters, tedders and balers means that we don't have to box quite as clever as we used to. The crucial techniques used to make hay from grass – such as when to 'ted' or when to 'turn', when to 'break out the swathe' or when to 'row up' – have less importance today than they had in the age before the internal combustion engine. And the 'windrows', 'cocks' and 'ricks' that played such a vital part in the managing and collecting of the finished product have been replaced by polythene and the vast balers that can cover hundreds of acres in a day.

It was in the original craft of haymaking that the haycock played such an important role. A cone-shaped pile of hay that is left in the field until it's dry enough to store, the haycock was the primary weapon with which the haymaker of old outfoxed the squall, sidestepped the downpour and ensured that the precious hay byre didn't go unstocked in preparation for the long winter months ahead. A well-built haycock sheds water, and the secret is not just to pile up the cut material and hope it doesn't get too wet, but to place each forkful of hay, much like building blocks in a wall, onto a firmly created base and build the body up, ensuring that each block overlaps with the course below to hold it fast. When the appropriate height has been reached, the cock-builder works inwards to create a conical roof. Then, the important and oft-forgotten part of the process: the combing down of the cock with the fork or rake, and the drawing of each stem on the surface in the same direction to improve water run-off. A good haycock builder will leave enough space on top to place a cap of combed-out material, and will also work the sides in such a way that they taper in towards the base. This creates an eavesdrip that further protects the body of the haycock from water.

Haycocks can be used at any point in the haymaking process. Obviously, if you find yourself immediately 'cocking up' after a cut, you've made a bad call on the weather. So getting the initial window of a day without rain is crucial. My advice in any reading of the weather is to climb up to the nearest high place and face into the wind. Squint, scratch your chin and glance at your watch (for effect), then look to see if there are any nasty rain clouds in the distance. You're most likely to find yourself building a cock at the end of the day to protect your hay from overnight showers or a particularly heavy dew. They are also a crucial emergency measure: at the first sign of heavy cloud, rush out and cock up to avoid losing your crop. A well-built haycock will withstand a series of moderate showers and, when the weather is looking up, can be broken out and spread about for the grass to cook off some more. With cocking among your arsenal of techniques, you can make hay in two or three windows of sun rather than requiring a continuous four- or five-day spell of dry weather. So it's a technique well suited to western Britain, where the frequent squalls and showers that come in off the Atlantic Ocean can test even the most experienced haymaker.

I don't know if the drawn-up cut grass from the trunk road embankments at Ilminster were purposely constructed. They might have just been tidily collected piles of grass destined for the council green waste site. But they fired up my vivid imagination and, most importantly, they brought home to me how haymaking is not just a simple process but one that requires the successful juggling of an inordinate number of variables – a true *cræft* – an ineffable ability to turn nature's gifts of sun, wind, rain and the reproductive properties of plants into a source of fuel for livestock. Today, more often than not, we consider the traditional production of a scythe or a pitchfork as the 'craft'; but it is the correct use of these implements in the field that represents the *cræft* – the longer trajectory of production and use within a wider socio-economic context.

3

STICKS AND STONES

IN APRIL 1997, at the snooker world championship held at the Crucible Theatre in Sheffield, Ronnie O'Sullivan stepped up to the table to play a frame in what was expected to be a routine victory in his first-round match against Mick Price. What happened in the next five minutes and twenty seconds sent shock waves through the world of snooker and ripples of respect through the wider world of professional sport. To the uninitiated, there is a sequence of thirty-six balls that must be potted in order to achieve the highest score possible in a frame: 147 – what aficionados call a 'maximum break'. Up until 1997, this had been achieved in official competition snooker on a handful of occasions, in a sport that had effectively turned professional in the late 1960s. It was only a matter of time before the gifted O'Sullivan scored his first competition 147, but it was the manner in which he did it that created such a stir. As he glided around the table he played with a pace and confidence that belied his twenty-one years. A man at one with the stick in his hands and in a trancelike engagement with his art, he was demonstrably thinking four or five shots ahead and, in playing with such fluidity of movement, O'Sullivan had found a new zone within which the game could be played.

It may seem crude, but to put the achievement into context, it can be compared on pure financial terms with other sports. For a frame that lasted a mere 320 seconds, O'Sullivan was awarded bonus prize money of £165,000. Few can brag that they've ever earned £515.63 per second for the work they do – especially at such a tender age. At its most basic, he makes his money with a length of polished wood and a lump of chalk. For many people, earnings aside, O'Sullivan's feat ranks among the very best sporting achievements in the world. But for me, it's a celebration of mankind's perfection at stick usage: a poetically beautiful combination of craft, genius, nerve and swagger.

I'M BEGINNING THIS chapter with sticks because it's probably where the story of craft begins – the point at which our very distant ancestors progressed from animalistic existences to lives materially enhanced by the objects around them. The transition is most notoriously depicted in the 'Dawn of Man' sequence in Stanley Kubrick's *2001: A Space Odyssey* when, in a moment of epiphany, an ape holds aloft the bone he has just used to pulverise to death the leader of a rival tribe before casting it up into the sky. It's unfortunate that my example of humankind's breakthrough moment in the evolution towards tool use occurred in such violent consequences. Kubrick's objective was undoubtedly to comment on what drives technological change, and how using sticks to fight each other was instrumental in the development of human societies. But I suspect they played a more mundane role in our evolutionary journey before they were systematically used for brutalising fellow members of the species. Even with the orchestral soundtrack provided by the climactic opening bars of Richard Strauss's *Also sprach Zarathustra* the sequence would have lacked a certain potency if Kubrick's ape had used a stick to knock an apple off a tree. Whichever way you choose to depict this defining moment in the human story, the successful use of a stick in those primeval times would undoubtedly have brought fame and fortune.

Over three million years on, that rule still applies today in many cultural circumstances. Technically speaking, snooker belongs to the world of sport not craft. But looking at sport as an extension of the physical prowess it took to compete – and to be the best – then I have no issue with extending the notion of craft to the work of sportspeople – especially those who employ sticks. Tennis players, cricketers, snooker players and golfers, to name just a few practitioners, all wield sticks-of-sorts in a skilful way. And so we arrive back at Ronnie O'Sullivan and a trajectory of hominid stick usage that takes us from its perceived beginnings, as imagined through Kubrick's ape, to its zenith, the Crucible Theatre, Sheffield, April 1997, and the fastest maximum break in history.

Yet Kubrick could just as well have substituted the bone with a stone, and in doing so may well have been more accurate in his portrayal of seminal tool adoption. While stones and bones survive in the archaeological record of early prehistory, it's hard to know, unless there are obvious diagnostic signs of wear or modification, if a bone has been used for adapted purposes. Wooden sticks present an even more challenging situation by virtue of the fact that, unless suspended in the extreme environmental conditions of desiccation or saturation, they decompose and turn to dust. Stones, on the other hand, survive the ravages of time and make it abundantly clear to us when they have been refashioned or altered by the human hand. Thus they provide the earliest evidence for the human use of tools and have come to define the way we understand the development of human societies from around three to four million years ago until at least the Bronze Age (*c.*2500–800 BC).

The byword in archaeology for stones is '-lith', ultimately deriving from the Greek λίθος, meaning 'stone'. It is on the basis of a stone-tool typology that we have been able to establish a chronology for the Stone Age. From the Palaeolithic (the 'old' Stone Age) through the Mesolithic ('middle') to the Neolithic ('new'), stone tools become progressively more complex. It's a story that begins around three million years ago at a place called Olduvai Gorge on the Serengeti Plains of Tanzania, and includes the work of the British-Kenyan palaeoanthropologists and archaeologists Mary and Louis Leakey and their excavations in the 1950s. Here skeletal remains of *Australopithecus*, an early apelike hominid, were recovered, alongside associated assemblages of worked stone. These early tools are usually labelled pebble or cobble tools because they appear to have been struck only enough times to create a single sharp edge. So these early tools were really very basic. Yet for *Australopithecus*, whose diet comprised scavenged meat, they were undoubtedly a step up from pulling apart a carcass with their bare hands, and allowed for the scoring of the hide, severing flesh and the breaking and crushing of bones to release marrow. This small but significant step would lead to increased protein consumption and thus had a long-term evolutionary impact.

Then, around 1.9 million years ago, *Homo habilis* arrives on the archaeological scene, shortly followed, at around 1.2 million years ago, by *Homo erectus*. We now start to talk of hominins – members of the human clade – defined against the wider classification of hominid, which contained more apelike members of the genus, such as *Australopithecus africanus*. We tend to call the worked flints from this period Acheulean, after an archaeological site located at St Acheul on the outskirts of Amiens, northern France. Here, in the nineteenth century, a number of what were termed hand axes were recovered from the gravel river terraces of the Somme region. In some ways, it was at this point that the Stone Age was born, as the incontrovertible evidence of stones that had been altered by human endeavour, associated with geological deposits of known age, forced a reconsideration of the traditional biblical narrative of how we were created.

Acheulean hand axes are beautiful artefacts to behold. For my first ever lecture on archaeological illustration at the Institute of Archaeology in London I had to make a technical drawing of one of these axes. As I turned it over in my hands I marvelled at its epic journey through time. These beautifully worked flints show obvious signs of repeated striking to work a core down to a finished axe that has sharp edges on two sides converging on a tip, but with a 'hold' at its base or distal end. What is so mesmerising about them is that, written into their fracture lines, one can see the consciously made decisions and the cognitive processes of design as the lower Palaeolithic knapper conceived the desired shape and form. Here was something truly 'human'. The term hand axe is, however, probably a bit of a misnomer.

On an experimental trip to the dense woodlands of the Sussex Weald in the late 1990s, a few friends and I decided to see if we could fell a tree with crudely made versions of our own. Proponents of the original Palaeolithic Acheulean school would have undoubtedly winced at the standard of our replications, hurriedly knocked out in the back garden of a terraced house in Haringey one hot summer day, before catching the train down to Sussex. But ours certainly had sharp edges, and some very willing experimental archaeologists happy to spend a weekend hacking the trunk of a tree with them. In truth, the endeavour lasted little more than a few hours. Our arms and wrists quickly tired, joints started to seize and swell, and regularly swapping hands only served to spread the agony. So traumatised were the bones and muscles in our wrists that we could barely lift the consolatory pints to our lips at the local country pub that evening. Sucking ale through brightly coloured straws, we all concluded that we should probably view the hand axe as an Acheulean Swiss Army Knife or Leatherman, a kind of multi-purpose tool. We are now encouraged by the experts to envisage hand axes as having a range of functions, including basic butchery, breaking, chopping, scraping, crushing and digging, as well as being a form of currency.

The end of the Acheulean industry broadly overlaps with

the emergence of *Homo neanderthalensis* and *Homo sapiens* between 100,000 and 125,000 years ago. With this ushering in of the middle Palaeolithic comes a much more developed attitude towards tool production and an increased sophistication in terms of social order. While artistic and symbolic representations were perhaps beyond their consciousness, their burial practices and other rituals are evidence of a capacity for abstract thought and a degree of self-awareness. Stone tools of this period are often referred to as Mousterian, after the type site of Le Moustier in the Dordogne where some of the earliest most complete assemblages were found. Hand axes continued to be standard fare, but the period is also characterised by what we call 'scrapers' – small hand-held flints around the blunt side of which the index finger is wrapped to create an effective cutting tool. These scrapers were almost certainly used in the preparation of hides, and the remarkable resilience of both *neanderthalensis* and *sapiens* in the face of climatic variation suggests that more sophisticated protective clothing was being produced.

Despite this, *Homo neanderthalensis* is thought to have died out at around 40,000 BC, at the beginning of what was an extremely cold period for Europe. From here on, from the upper Palaeolithic into the Mesolithic, stone tool manufacture is characterised by much variation, innovation and rapid development. Not only were the stone tools more sophisticated but they were also used to create bone tools such as awls and needles. Both suggest further developments in clothing and the likelihood that composite garments were stitched together for a tighter and more ergonomic fit. I often say to friends in the bespoke tailoring trade that it is to *Homo sapiens* of the upper Palaeolithic that their craft owes its greatest debt. Without those needles and custom-fit lines we might never as a species have survived that cold snap.

In the stone tool record local traditions are also evident, a sure sign that *Homo sapiens* had the capacity to adapt production to local environmental conditions. We can almost start to talk of 'cultures' because of the variation in what are termed blade flakes – tools made from the flakes knocked from the core. It seems surprising that it took so long to work out that the bits flying off the core were as sharp as the core itself. Back in our garden in Haringey it took a matter of minutes before barefoot flatmates found just how sharp a casual scattering of waste flints – or 'debitage' as it's technically known – could be. But the innovation wasn't that tools were being made from off-cast flint but rather that the core was prepared in this way in order to purposely produce blade flakes from it. It's also obvious from analysis of the flakes, cores, platforms and what we call the 'bulb of percussion' that a whole range of fabrication techniques was being used. Indirect percussion (rather like the method of hammer and chisel), pressure flaking and soft-hammer percussion (with an antler, for example) all allowed the upper Palaeolithic knapper to create a vast variety of stone tools that supported a very sophisticated relationship with the natural world. If this short history has captured your imagination then I advise booking yourself onto an introductory flint-knapping course. It's easy to become lost in the immersive world of smacking stone on stone; it's an enormously therapeutic pastime, and one that allows you to connect with the innate *Homo sapiens* in yourself. There truly is no more authentic way of getting back to basics.

The key technological development in this period is the evidence for hafting – the fixing of a spearhead or arrowhead onto the end of a stick. The evidence comes not from the excavation of complete weapons – wooden shafts with blades attached – but from the shape of the worked flints and the presence of side and corner notches at their base. These indentations cut into the flints would provide purchase for a length of cord used to bind the blade to the stick. There has been a long-standing debate as to whether, as early as the Mousterian industry, projectile points were hafted, but recently excavations at Kathu Pan in South Africa have recovered a number of stone points whose tips exhibit fracture types that indicate impact rather than scraping and sawing. Furthermore, modifications near the base of these points were consistent with hafting. The scientific dating from the site proposes a date range centring around 500,000 years ago. This is an incredible 200,000 years earlier than conventionally thought and forces us to rethink man as hunter rather than hunted at a much earlier period in our evolutionary models. What is certain is that these crude attempts were nowhere near as sophisticated as the projectile points hafted in the Mesolithic, nor those still produced today by cultures in the Amazon rainforests of South America.

Hafting – the technological capacity to attach a stick to a stone – really is the point at which craft becomes cemented as an evolutionary option for the human species. The composite tool or utensil was born, and with it the capacity to make at a much more advanced level than before. That seminal moment of creating a weapon or tool is, in my opinion, a crucial coming together. It is an event that signals a new dawn in human technical advancement – effectively the creation of an extended limb – and one that is certainly well developed by the Mesolithic. Whether it began 500,000 or 300,000 years ago, I'd like to pick up the hafting story in its final days, somewhere in the 1950s, with my grandfather, the former golf-club maker.

OVER THE LAST couple of years I've given in to an urge; it's a subliminal longing and a deep genetic inheritance. I have started making sticks. It's addictive – and evidently in my blood.

I blame my grandfather on my father's side. He passed away when I was only two, so I never really knew him. He was born, lived and died in St Andrews – the home of golf – and he plied his trade in the golf-club-making industry. He was, therefore, a stick maker par excellence and a man of considerable skill. My father recalls how, as a boy, he would watch in awe as his father sat outside at the back of the house turning lengths of hazel, ash or willow into walking sticks. The scene was to repeat itself a generation later as I watched my own father sitting on the back step, whittling away in the evening sunlight. I can't remember a time when my dad didn't have a stick or two on the go, carving intricate figures into the handles and decorative patterns into the shafts. But I never once thought I would find myself in his place, daydreaming while my hands took over the work.

My grandfather lived a troubled life, one best illustrated by the story of his bike. During the winter Grandad used to cycle to the gasworks each week to pick up three sacks of coke – a by-product of the coal-gas industry – to burn on the home fire. He would put one sack on his pannier rack at the back of the bike, balance another on the handlebars and position the final one in the frame between his legs, then precariously pedal home. It was on one such journey, as he was cycling along the quayside, that Grandad's bike started to fail him. The rage – or *radge* as it's pronounced in Scotland – kicked in. Coke skittered across the quayside, the bike ended up in the drink and my grandfather stormed off to my great auntie Jean's house to lick his wounds. For years afterwards, Sandy Langlands's bike could be observed in the harbour bed, slowly immersing in the tidal sands of time. Occasionally re-exposed by the scouring action of a particularly harsh swell-tide, it became a familiar monument to sub-working-class living conditions in post-war Scotland. Heirlooms are in short supply from the commodity-starved Scottish side of my family. I know there is a handmade putter made by my grandfather somewhere among my father's possessions, and I hanker after it. But should it pass to either of my siblings, if I am to have anything to remember my Scottish ancestry by, I might have to settle for trying to extract that bike from its maritime grave.

I'm sure the reasons for my grandfather's frustration, his *radge*-like outburst and his hurling of his only means of transport into the harbour, lay in the pennilessness of his situation and the lack of work in his chosen profession. Before the Second World War Grandad had made his living as a golf-club maker. But war and mechanisation had taken their toll on the industry, and when the sport re-emerged in the 1950s as a pastime of the wealthy and the aspiring middle classes, demand for a set of clubs was met primarily through developments in machine technology. The age of the handcrafted golf club drew to a close, and with it my Grandad's livelihood. From then on, between long periods of unemployment, he was forced to turn his hand to all manner of menial tasks to make ends meet. Ultimately, it was because of this that he found himself sourcing a second-class fuel, on a rickety old bike, making his way back to his hungry children on a cold winter's evening.

Among the many manual skills required to make a golf club, one of my grandfather's specialisms was the whipping of the club head and handle to the shaft. It was a relatively simple process but one that was absolutely necessary to get right if the joint between the head and the shaft was not to fracture, open up under duress and eventually split apart. It involved taking a piece of cord and binding it around the joint as a means of reinforcing it. With one hand, pressure had to be kept on the cord at all times to keep maximum tension while the other hand slowly turned the head of the club to wind the cord on. The job was completed in such a way as to conceal the knot and give it a smooth finish, for appearance's sake but also so that no trailing cord was left to snag and weaken the bind. Although robbed of an arena within which to use his craft, Grandad passed down to his son the skill of whipping and, in turn, my father has passed it down to me.

I've had cause to use this inherited and versatile skill on a number of occasions, most recently in attempting to make a late medieval/early modern fishing rod. In a text called *The Treatyse of Fishing with an Angle*, dated to the late fifteenth century, a type of fishing rod made from hazel and a length of horsehair line is described in some detail. Hazel is a wonderful wood for creating fishing rods. Grown in the shade and in wet, humus-rich soil, it can attain dead-straight lengths of ten to twelve feet while remaining remarkably thin. It's flexible too, so it can take the strain of the fish as it tries to wriggle free of the hook. Where it falls short is in the weakness of the soft wood at its thinner end. This is prone to snapping under the slightest pressure. Therefore, a small wand of harder wood – such as blackthorn or applewood – needs splicing onto the hazel at a thickness where it's strong enough to take the joint. The two woods are bound together using the technique of whipping, which can also be used to create a grip on the butt end of the rod. By the time I'd finished, my fishing rod certainly looked the part. Pity then that I had neither the patience nor the skill to fish.

It was on the first occasion I had to use whipping that I realised just how ancient the technique was. I was trying to fasten a flint spearhead, which I'd spent a good two days producing from scratch, onto a shaft of ash that I'd bark-stripped with a flint scraper. I was feeling rather smug, for this was the kind of thing I used to do as a kid with schoolfriends during long hot summers spent in the spinneys, holts and coppices that border Pevensey Marsh in Sussex. Now, as a student in higher education, with an interest in experimental archaeology, I could dress it up as 'research' to justify the hours spent returning to the halcyon days of my youth. I was going for something with an upper Palaeolithic feel; something with which I could burst out from behind a hedge and take down a young caribou – or at least that's what I imagined.

I experimented with indirect percussion and pressure flaking in the process of making my arrowhead, and used commercial cordage for the whipping. I'd hoped to make my own string from the stems of nettle plants, but was rapidly running out of time – a common problem when trying to cram the Palaeolithic into a three-day weekend. As I sat in the dappled light of the woodland floor, intently concentrating on the binding process, it dawned on me just how far this skill, or technique, had travelled. It had passed through hundreds of thousands of generations, crossed continents, spanned epochs and fulfilled a multiplicity of functions on its way. And here I was, using it by virtue of the time taken by my father and by his father before him to pass it on. It told a story that was as important to our understanding of humanity as any written history, a tale of ordinary people who relied on such skills for their sustenance. When the machines took over this process in the golf-club-making industry of St Andrews, my grandad lost his livelihood and his source of pride, but we, the wider society, lost a direct and tangible link with our ancient ancestry.

I WOULD LOVE TO be able to tell the parallel evolutionary story of stick development since the lower Palaeolithic, for it's almost inconceivable that *Australopithecus*, *Homo habilis*, *erectus*, *neanderthalensis* and *sapiens* did not develop this technology to the same degree of sophistication as they had stone-tool technology. But because of wood's inability to survive in the archaeological record, it will for ever be a story that remains untold and one merely hypothesised by the daydreaming of experimental archaeologists such as myself. Even if we can't chart the development and diversity of its usage in primitive society, we should at least thank the stick, the stone's silent sister in the archaeological record, for the part it played in developing the cognitive processes of the human species.

It was in my own collection, an idiosyncratic curation of useful and interesting sticks, that I began to see just how central the stick had been to a wide range of industries in the rural economy. I hadn't purposely set out to collect so many sticks, but one day, as I was sorting through one of my toolsheds, the stack of long, thin sticklike implements stood out. Intrigued by the growing assemblage, I started rummaging through other sheds, as well as the garage and store cupboards in the house, in a bid to bring together as many sticks as possible. I decided that anything long-handled, whether or not it had an attachment on its end, should be accepted into my broad definition of stick.

Standing before me was an unwieldy collection of miscellaneous farming, household and gardening tools. There were the classics of the farming and gardening world: spades, shovels, forks, hoes, scythes, rakes and so on, along with some rarer and more specific sticks, of which my own personal favourite, the 'whin bruiser', stood out. The everyday household contingent was mostly staffed by an assortment of brooms, dusters and mops, as well as a bespoke stick used to open the loft hatch. There was also a mixed bag of sporting sticks: a cricket bat, my father-in-law's fishing rod, a hockey stick, some old badminton rackets and two pool cues. And then there were the sticks proper and a miscellany that doesn't fit into the previous three categories. Of the former, a selection of walking sticks provided the mainstay. But it was the miscellany that proved of most interest. There were a handful of willow-wood rails I'd made for a Sheila Maid

– a traditional clothes airer – and as replacement rails for some gate hurdles on which I used to grow peas in the garden. There were my chimney-sweep brushes and some drain rods. And a yoke I'd made out of a length of blackthorn wood that had grown around a fallen tree trunk in such a way as to make the perfect curved indentation to position around the neck and rest on the shoulders. On woodland or hedging jobs, I often used it to carry two heavy but evenly matched tool bags from the car, leaving my hands free to carry a third bag. There was an oar, lantern staff and a net-pole – happy memories of messing about on the river one summer. And what I called my mole spear – not because I used it to spear moles, but rather to spear the lawn to test for mole runs, assessing the lie of the battlefield before laying my traps. But by far and away my favourite two groups of sticks were my collection of shepherd's crooks and fruit pickers.

The shepherd's crooks were my sticks-of-dreams. They represented a desire I'd harboured, since my late teens, to become a hill sheep farmer. It's a dream that, as I approach my forties, is rapidly dissipating but one that on occasions still burns strong – strong enough for me to have spent a considerable amount of time crafting my own series of crooks. I'm not completely alien to sheep farming as I've spent at least three years looking after small flocks for various BBC farm series. But what I really wanted to do was to see a flock of three hundred or so sheep through the hill-farm year – to really get a feel for that lonely existence on the hills.

It was during my first stint as a budding historical farmer that I came to appreciate that the shepherd's crook was more than just an obligatory accessory for the nativity play shepherd – and that different types of crook could be used for different parts of the sheep-farming year. Our historical farm was stationed on the border between Wales and England – both great sheep-farming nations – occupying a small complex of buildings at the head of a steep-sided valley. Above us were open summer pastures of moorland – the 'commons' – and below us, the rich winter grazings of the valley. The head of the re-enactment group charged with stocking our farm had assigned us a rather modest flock of five Cotswold Lions, a majestic breed from central England, somewhat ill fitted to hill-farm territory but stupendous in its production of beautiful long white fleece. Our hill farm was to be set in the early part of the seventeenth century, a period of Britain's agricultural history when cloth production took priority over all other forms of farming. Wool was everything, and if our series was to be historically accurate we had to reflect that.

Well, it didn't take long to realise that something was up. Not only were our four ewes coming into heat, when they should have been showing signs of early pregnancy, they were also in fantastic condition. Both were signs that their wombs were singularly devoid of lambs. On the basis that all four were in such condition, the finger was pointed squarely at Cyril, our ram. In a moment of candidness, the head of the re-enactment group confessed that a couple of years ago Cyril had got so wrapped up in bramble he found himself trapped and, exhausted by the struggle, collapsed to the ground. He was found three days later, alive, but unfortunately for Cyril (and us, as it turned out) the hard winds blowing in from the Welsh mountains had frozen his precious testicles to the ground. Clearly, the experience had rendered him impotent.

Of course, this all caused a huge storm for the television production. Having invested so much camera time in the 'sheep-were-everything-in-the-early-seventeenth-century-rural-economy' storyline, we now needed to see some lambs being delivered. After much wrangling, it was decided that a few pregnant ewes would be purchased and seamlessly slotted into the flock to then deliver the much-desired lambing scenes for our spring episode. The viewer at home would be none the wiser.

When the stock wagon arrived, loaded with our new sheep, I rushed down the hill to where it was waiting, only to find Peter, my fellow presenter, crying with laughter as he peered through the slats of the trailer. As I came alongside him I gazed at four healthy and heavily pregnant sheep. But they were black. All four of them.

The Welsh Mountain Black is a particularly hardy breed of sheep well suited to the challenging terrain of the Welsh mountains. In these early years of sheep husbandry it taught me that there is an incredible variation of characteristics among different breeds of sheep. Compared to the Cotswold Lion, the Mountain Black is almost an entirely distinct breed of animal. Whereas the Lion is tall, long and covered all over with lengthy, white, shaggy wool, it isn't anywhere near as nimble and hardy as the Mountain Black. The wool so prized by the lowland farmers of central England had proved Cyril's downfall amid the challenging vegetation on the foothills of the Welsh mountains – with his heavy, meat-producing frame and muscle structure he'd been unable to wrestle free of his bramble bonds. We also discovered that the Mountain Black was an altogether different prospect for the shepherd too. With the Cotswolds, they'd been so bulky and sluggish they were easy to herd across the common and guide around the farmstead. With the Mountain Blacks, it was like trying to herd wild deer. We simply couldn't get close to them.

Having released them from the stock wagon, we spent the best part of an afternoon running around the common, cautiously trying to outmanoeuvre them before finally admitting defeat. With them being so heavily pregnant, it would have been dangerous to try too hard, and for too long, to get them to do our bidding. The fact is, they would probably lamb just as safely out on the common as they would in our lambing pens. And in any case, on the basis that the sheep were a completely different

72

colour, the director had somewhat given up on trying to stitch together footage of Cotswolds Lions struggling through a hard winter with footage of Mountain Blacks lambing in spring.

But it was in my dealings with the local sheep farmer, whose dog we'd been forced to draft in to help round up our flock of four black sheep, that I began to see the benefits of having a range of shepherd's crooks. His farm was just over the hill – a ramshackle affair of barns, lean-tos, prefabs and an old cottage – and he, like us, was in the middle of lambing season. I learned a few handy tricks from this chap. As I gazed around his lambing stalls I marvelled at the ingenious secondary use of crate and pallet wood to create individual pens. In one corner I observed a sheep with its head stuck in what looked like vertical stocks. It could move its head up and down to get to hay and water, but not from side to side. The old shepherd explained to me that a lamb had been orphaned the previous night and that this ewe, already with one lamb, was being persuaded into fostering the other. After a few days it would become accustomed to the sound, smell and feel of the orphaned lamb and could be trusted not to spurn it. At a later point in the year I observed a paddock with two mature ewes with rather thick necklaces of twisted hay. These stubborn ewes, I was told, had taken a disliking to one another in the field and were almost incessantly butting and harassing each other. These edible Elizabethan-style ruffs of hay were the only source of food in the pen, so if the battling ewes wanted to feed they had to get up close and nuzzle, ultimately developing a bond of familiarity.

It was in the use of the crook that this experienced shepherd was most insightful. For separating out his sheep he had a very traditional-looking crook, the head of which was a ram's horn carved in such a way as to create a smooth but open loop. There's a point after you get a flock of sheep into the corner of a pen when they all cram into each other, ears pinned back – a wall of sheep's arses. It's nearly always the case that the sheep you want to separate out are the ones furthest away from you, and to get to them you have to wade through a sea of ovine backsides. Here the crook comes into its own and is judiciously utilised in hooking around the neck and drawing the desired beast towards you. When rural communities communally sent their flocks out to summer pastures on the mountains, heaths and marshes, the gathering up and separating out of individual flocks on returning to the winter fields was a big job, and one rendered impossible without the crook. It was a vital extension of the shepherd's arm, and mastery of it was an indispensable requirement if flocks were to be efficiently separated out without needless stress to the animals.

There is another type of crook – the leg crook, or 'cleek' – that finds employment at a different time of year. On this crook the loop is smaller, just wide enough to place around the hind leg of a lamb, and curves gently out into a spur that acts as a guide, funnelling the leg down into the loop as the crook is drawn back towards the shepherd. The cleek comes into its own at lambing time. As a beginner, it's easy to get complacent around newborn lambs. For one- and two-day-olds you can just step into the pen and pick them up. But over the course of a few weeks they develop lightning quick reactions, and before you know it they're outstripping you, darting around and sending you in all sorts of directions. It's at this developmental stage that the shepherd gets paid his lambing wage, at the age when the lambs can safely look after themselves. But to get them to this age the shepherd must keep a watchful eye and tend to them at every opportunity. The leg crook buys you those extra few days – perhaps a week – when you can get close enough to hook them and handle them to undertake necessary assessments of condition, ear-clipping, marking, castration and, more recently, administer vaccinations and deworming medication. More time spent handling them also diminishes their 'flight zone', meaning that the more time they spend in your arms and come away safely, the less distance they will habitually place between you and them. So the leg crook really was the shepherd's moneymaker. It gave him the vital edge needed to maximise conversion rates in the flock, over a six- to eight-week period, from pregnant ewes to mothers with lambs.

Any traditional shepherd would be lost without his sheep dog to do his running and rounding up, but he would be equally adrift without his suite of crooks to handle the flock at crucial times of the year. It was the development and use of these sticks that facilitated good husbandry – the ability to handle your livestock with care and efficiency. To tend them. And here we find ourselves at the transition from hunter-gatherer to farmer, to the domestication of livestock, and a significant chapter in the human story that we know began around ten thousand years ago. Those early farmers would have had in their possession a range of modified sticks to aid them in their handling of their flocks. Unlike the stones used to slaughter and butcher, these sticks haven't survived. But we must surely hypothesise their existence and the fundamental role they played in the development of agrarian societies into the Neolithic and beyond. Today, you can go to an agricultural supplier and choose from a limited range of aluminium crooks. They aren't pretty, but they're incredibly light and indestructible. Amazing to think that a design so simple still finds function on farms today – a stick that has enabled the handcrafting of Britain's famously varied breeds of sheep.

ANOTHER OF MY favourite sticks is a ten-foot length of ash with three branches protruding from the end to create a three-pronged crutch. From the moment I clapped eyes on this extraordinary length of timber, minding its own business, growing from an ash stool (stump) in the hanging woods above my cottage, I knew immediately what I had in mind for it. Before you could say 'by hook or by crook', I'd felled it and raced back to the cottage to press it straight into action. There's an enormous Bramley apple tree just across the lane at the top end of the garden. The only traffic that uses the lane is agricultural, and to enable the vast tractors and harvesters to pass throughout the year the tree is regularly flayed back to a height of about eight feet. It suffers the same fate on the other side, where it overhangs an arable field, and as a consequence it yields very few apples within easy picking reach. During the early years, my raids on this tree had been primarily for cider-making purposes. The sourness of the Bramley complemented beautifully the other two sweeter varieties of apple I had growing in the garden, and together served to produce a medium sweet cider with a dry aftertaste. But there is only so much cider a man can drink – and still function – and over time my preference for fermenting turned to keeping. This is probably the simplest way of ensuring that you have a supply of apples for the most part of the year. Kept in the right conditions, certain varieties of apple will keep for a remarkable length of time without the aid of preserving agents.

One particularly good keeper is a heritage variety of apple called the Flower of Kent, and I witnessed first hand just how well this fruit can survive through the winter months and remain edible into spring. While working as an archaeologist I was occasionally called on to record ancient vernacular farm buildings that were due for modernisation. This work took me all over the south of England, staying in obscure B & B accommodations, and allowed me to see the hidden countryside of Sussex, Kent and Hampshire. It was while wandering to work one day on just such a trip that I passed an orchard of heritage varieties of apples, all looking pretty ripe for the picking. For my sins, I am an inveterate scrumper, a habit I picked up as a boy when I was in France. But I'm not a greedy scrumper. I made my selection carefully and plucked four of the largest, sun-soaked apples I could find before making my escape. One I dispatched on the remaining stretch of my journey to work, two I handed to my colleagues, and the fourth I placed in my waist pocket. Later that day, while squeezing through the attic timbers of a stone-tiled barn, I found the apple substantially hindering my progress. I took it out of my pocket, placed it between the angle of the crown post and tie beam and continued with my work, resolving to pick it up on the way back through the timbers. Of course, I forgot all about it until April the following spring, when we returned to the barn to complete the job. I was amazed to rediscover the apple exactly where I'd left it. I picked it up and turned it around slowly in my hand. It was slightly smaller, withered and had lost a bit of colour but, to all intents and purposes, it was a perfectly preserved apple. Instinct took over and I bit into it. Drier, a bit stale, perhaps not as sweet as it could have been but very definitely edible. In fact, simmered slowly in a pan of shallow water, it would have made a very fine apple sauce.

It was obvious that the conditions this apple had been kept in had preserved it. The stone tiles had regulated its temperature throughout the winter. A light breeze had kept it dry, and it was high enough to be out of reach of vermin. No preserving vinegar, sugar, yeast or salt required. But the key to this apple's keeping also lay in the way it had been picked and gently carried away. Because it hadn't been shaken from the tree and sent crashing to the ground, or plucked by hand and tossed into a basket, it hadn't bruised. It's in the bruising of a fruit that its protective membrane, its skin, is weakened, causing an opening into which fungus spores can penetrate and the apple to mould. Put simply, look after your apples when picking and they will look after you through the winter months.

You therefore need to adopt a slightly different picking method, and this is where my bespoke ten-foot apple-picker came in handy. It was versatile and could be used as a 'panking pole' – the west of England name for a long stick designed specifically for shaking apples from a tree. But the crutch created by the prongs also allows you to cradle the individual apple, gently twist the pole to sever it from its bough and then bring it down to be carefully placed in a tray for storage. Actually, my method is a bit time-consuming, and I have since seen similar care taken when picking apples in central France but at a much greater speed. On this occasion, the pickers worked in pairs. The younger of the two would climb up the tree and, using a long, thin stick with a stunted end, would jab sharply at where the stem of the apple connects with the branch to free the ripened fruit. Beneath, the second picker would wait, cigarette hanging loosely at the side of his mouth and an outstretched neckerchief between his hands, catching the individual apples as they fell and delicately placing them in the basket on his arm. He concentrated intently as his colleague worked his way around the tree with remarkable dexterity. For the method to be of economic viability, a high level of accuracy, speed and fluidity was required. I strongly suspect that Ronnie O'Sullivan would be a dab hand.

4

GRENJAÐARSTAÐUR

AS THE PLANE banked around the bay to approach the airfield, the landscape of northern Iceland opened up before me. I looked across a rugged and hostile place clinging to the edge of northern Europe, wincing at the prospect of the cold southerly arctic winds. My destination was the small fishing town of Akureyri, the district capital of the Northeastern Region, and as we'd flown up from Reykjavik, I'd spent much of the journey craning my neck, peering out of the small cabin window at the brooding world below me. The journey had taken in mountains, glaciers and deserts of volcanic basalt rock, but as we neared the coastal valleys vast expanses of heaths and moors opened up, periodically studded with tiny enclosed farmsteads, each set within a hollow of small fields and connected to the road network by long sinuous tracks. It seemed to me a landscape in complete contrast with that of my homeland. As we'd flown out of London Heathrow we had left behind a densely settled place, remorselessly partitioned with every single parcel of land defined, marked out and abutted by another. It was only as we'd hit the Midlands and further north that areas of open land began to appear. But even these were boxed in and encroached on – islands of apparent wilderness within an otherwise tamed and incarcerated landscape.

In the skein of lines that webbed out beneath me the piecemeal abduction of Britain's countryside could be discerned. Processes of reclamation were identifiable in the sweeping curves of hedgebank and lane. Internally subdivided, these pioneering intakes of domestication themselves became subjected to a ceaseless breaking up, apportioned and doled out as shares. Piercing arterial roads haemorrhaged pockets of development that had in turn swelled outwards, tumour-like, so that their extremities kissed those of the neighbouring villages through the thin membrane of the pervasive edgelands. Towns became cities, cities became conurbations, and between each the ever-diminishing flame of the natural world flickered perilously in the winds of change.

But in Iceland it was the wilderness that dominated. Stretching out as far as the eye could see, each terrain type seemed to merge with the next and then continue unbroken to the foothills and the mountains beyond. There were no sharp lines. Even the river meanders seemed marked by the smudges of gravel terraces and sprawling mudflats before filtering out into the sea through fuzzy estuarine silts. Only the highways were evidence of human interference as they purposefully carved their way with little or no regard for slope or contour. Farmsteads seemed to have landed like droplets from trackways spurred off the main roads, but were entirely surrounded by the unrelenting vastness of the wilds. There were limited configurations of house, barn and outhouse, and the surrounding fields rarely stretched to anything more than an in-field and an out-field. These were the farms of the *Landnámabók*. Drafted in the tenth century, and with a literal meaning of 'Land-naming book', this ancient manuscript is thought to reliably describe the colonisation of Iceland by Viking settlers in the ninth century. Aside from the presence of a few Irish monks, rumoured in some of the Icelandic sagas, this land-naming represented the first serious attempt to populate the harsh and inhospitable island around 870 miles due west from the Viking homelands in Scandinavia. The farms that were set out twelve hundred years ago are largely the farms that are still worked by the Icelandic people today.

This was my second visit to Iceland. On the first occasion, in 2005, I'd been travelling with a group of archaeologists researching how the people of the early medieval period interacted with the space around them. Our themes were governance, the practice of assembly and the role played by the Church in social cohesion. We'd visited numerous important church sites during our brief stay and had travelled to the heart of the island to visit the *Althing*, the main meeting place for the Icelandic people in the medieval and later periods. The trip had proved fascinating because in colonising an unoccupied landscape, the Vikings had essentially set out the ideal of their own early medieval settlement hierarchy. With no previous configuration of prehistoric or classical civilisation to inherit from, they were effectively starting with a blank canvas. This presented itself as the perfect mirror to reflect back to the other areas of early medieval Europe our group was collectively studying – particularly those where Scandinavian influences were strong.

I jumped at the chance to return to Iceland in May 2012 as part of a similar research project with the same bunch of intrepid archaeologists. Our mission on this occasion was to examine the relationship between definitions of local and supra-local identity, or, in layperson's terms, to explore the negotiated position in society between ordinary people and those much further up the food chain – their kings, lords and bishops – and how this is reflected in the structure of landscape. As a whole, the group's list of research questions revolved around issues of polity, neighbourhood, identity and power. But as I gazed out of the cabin window at the barrenness below me, a very basic research question, and one more directly concerned with the human condition, reared up in my mind: how did people actually *live* here?

It was a question that kept cropping up throughout our whistle-stop tour of the Northeastern Region, and was most pronounced during a visit to a series of abandoned farmsteads strung out along the Laxá river as it carved its way from the mountains of Ódáðahraun to the bay of Skjálfandi. The sites had been archaeologically sampled in preceding decades, and ash layers deposited in the stratigraphy by dated volcanic events allowed for a relatively straightforward sequence of occupation to be tightly dated. These were *Landnáma* farms, with the earliest phases dating to the late ninth century. Today, all that existed was a series of low banks and mounds covered with a coarse grass. Under closer inspection of this uneven ground, the lower courses of walls could be identified and the broad outline of rectilinear buildings made out. They were small settlements with two or three main farm buildings surrounded by a series of indefinable ancillary buildings. The excavations had apparently revealed an eclectic range of burial practices, evidence of the cultural liminality of this place, but the presence of structures interpreted as churches – and dated to a phase some two hundred years before the formal national adoption of Christianity – reminded us that religious conversion was not always a top-down imposition. The farms were evenly spaced along the valley, at a midpoint between the river to the east and the summit of a low-lying ridge to the west. The wind blew harshly from the north and we huddled round, collars drawn up to our ears, as we listened intently to our guide.

The high point of these farmsteads was around the twelfth and thirteenth centuries, before a steady decline – perhaps as a result of climatic changes – gave way to a phase of crude shepherd stalls and summer sheilings in the eighteenth century, followed by final abandonment at some point in the nineteenth century. As I looked around at the unbroken sweep of the valley floor from river to ridge, I considered what it was the people of these settlements actually farmed. The line of the in-field and out-field, the traditional means by which livestock and crops were managed on primitive farms, was only just discernible under a carpet of reinvading sedge, rush and coarse grass. Sheep must have been important – but you don't get by on wool and meat alone. Our guide informed us that some archaeological evidence from the environmental sampling of the sites suggested that barley had been grown, but the long-term success of this crop was in question because the climate was considered too cold to provide enough of a growing season for the plants to reach maturity and yield grain.

As the party worked its way back along the valley to our transportation – a fleet of 4x4s – we discussed the interactions these farms must have had with each other and with other parts of the island for both cultural and economic purposes. What trade networks were in place in order to maintain a varied diet? What gatherings helped to bond communities in such a dispersed settlement pattern? How did they perceive their position in relation to southern Icelanders and the wider world? And at what scale did their identity lie with the valley, the region and the nation? For me, though, I still harboured a desire to explore the fundamental issue of existence: how did they *live* out here?

The answer to my question lay in Grenjaðarstaður. Perched on a low rise overlooking a fertile plain circumscribed by a bow of the Laxá was an ancient farmstead. But this settlement wasn't deserted like the ones further up the valley. The presence of a modern working farm set back about a third of a mile from the traditional farmstead indicated that this site had continued as a focus for agricultural activity. It occurred to me that if I was to answer my question I was going to do it here at a settlement that continued to carve out a successful existence. Perhaps the abandoned farmsteads further up the valley had been the wrong place to ask the question precisely because they were abandoned, and that, ultimately, people hadn't managed to *live* there. Perhaps the story of *life* – existence – was one with a slightly longer-term perspective. The occupants of these farmsteads had spent the best part of a thousand years clinging on, fulfilling the vision of the original *Landnáma* Vikings, but in the end they had relented in the face of the latent hostility of the place. But at Grenjaðarstaður things were different. Not only had they survived here – *lived* here – but they had done it with style.

The historic core of Grenjaðarstaður was a conglomeration of buildings all grouped together so as to give the impression of a single building. A row of two-storey gabled timber frontages painted a brilliant white faced the east, while the western end was bounded by a large two-storey timber hall set lengthways so that its gables faced north and south. The area between these definable buildings was infilled with single-storey buildings of all shapes and sizes. The visible external walls were of drystone construction consisting of volcanic basalt rock bonded with earth, but what was remarkable about this composite structure was that it was almost entirely covered with turf. The occasional dormer window peeked out through the dense sward, and in a handful of places a chimney flue protruded from the ridgelines. Standing back and taking in the whole complex, it was tempting to imagine that it had emerged out of the ground – in imitation of the volcanic volatility of the Icelandic landscape – forcing its way upwards from the earth in an eruption of human resilience. And now, as a museum to Icelandic rural life, it lay dormant, draped as it was with a carpet of green grass, shimmering in the arctic winds.

Grenjaðarstaður was the kind of museum for which I have a passion. Delightfully free of signage, display boards and information panels, it was up to the objects to speak for themselves. Entering through one of the front gable doors, I was immediately struck by the soundproofing qualities of the walls. While the basalt rock provided structural integrity, sods of cut turf stacked in herringbone fashion provided insulation from the howling winds and the biting cold. Beautifully hand-sawn timber panelling lined the interiors of the rooms dating from the eighteenth and nineteenth centuries, but the further into the maze of rooms and corridors I crept, the more I found myself slipping back in time, into a more primitive arrangement of space. The buildings told the story of Icelandic life from the earliest days of settlement right up to the later decades of the nineteenth century. Each room contained the craft objects and utensils of the time, and as I wandered around, in awe at the range of implements and craft practices, I snapped away with my camera, recording as any good archaeologist would the contexts of the artefacts I had uncovered. The digital reel I produced was littered with the kind of photographs you see in the lifestyle pages of a Sunday supplement; a photo shoot at a carefully manicured rural retreat, featuring the life's work of a style guru creating a certain authentic and earthy feel. Except here, craft was a necessity not a choice.

And it was thus a place of resounding beauty. As far as I could discern, there was not a single craft necessary for life in this harsh environment that wasn't represented in one way or another. As I moved from room to room, I registered the various areas of historical Icelandic life that were represented by implement and artefact alike. In the first reception rooms saddles, bridles and harnesses were evidence of the use of sledges, carts and ploughs, the means by which flocks of sheep and herds of the iconic Icelandic pony were husbanded. Subsequent rooms were packed with carpentry tools of all types: saws, drawknives, gouges, chisels, clamps, drill bits, augers, mallets, vices, shaves, windlasses and various other pieces of kit were all testimony to how important timber was to the farm economy. In the corridors were stationed lamps that would have burned brightly with the whale oil Iceland was so famous for. Each room, irrespective of function, was bedecked with skilfully made baskets, chests and trunks of all sizes.

Items of general function, such as ladders, spades, forks, rope, pulleys, spikes, axes and knives, were everywhere, and among these were positioned some of the tools more specialist in their dealing with the Icelandic landscape. The turf spade and peat cutters, for example, enabled the settler to produce building blocks for the house and fuel for the fire. Icelanders have never survived on farming alone and the display of shotguns, spears, harpoons, stone net weights and fish hooks was a timely reminder that of the few mammals inhabiting the waters around this desolate island among the most populous are whales and seals. The Laxá too was famous for its rich salmon stocks, and I imagined how important the rich bounty of fish arriving every spring to spawn would have been to this remote community.

Meat, bone, oil and fish eggs would all have been important, but the pelts of seals were one of the most crucial materials available to the Icelandic farmer-fishermen. Thick hides of mature bull seal would provide ideal protection against the natural elements for the crew of a whaleboat, and the white coat of pup seals, the finest and softest of furs, would have been used for the linings of tunic and breaches. Carved moulds for pelted stockings and gloves were hung from the rafters in one of the attic rooms, and rolled hides and pelts were suspended on string ropes as a preventative measure against invasive rodents. There was a full assembly of wool-working equipment too, from shears and hand carders to spinning wheels and a loom. Hand-cranked sewing machines allowed the cloth woven on site to be stitched together, and although cloth would also have been brought in, from the range of textile and pelt-working equipment on show it would have been possible to deck out an Icelander from head to foot with garments made from locally sourced materials and labour.

There were two kitchens on display. The first was a reconstruction of a medieval-style hall with a central open hearth. The walls of large basalt rock construction were exposed, and timber framing supported a roof of birch-wood brush thatch overlaid with heavy turves. A single open skylight allowed a shaft of sun in and a trail of smoke from the open hearth out. The structure of the hearth was remarkable. A large stone platform, perhaps five by ten feet across and rising to about knee-height, stood in the centre of the room. On top of this were large thick rectangular slabs of stone set at right angles to create four fire bays, leaving enough room at each end of the platform to take cauldrons and cast-iron pots off the heat. It was like a giant four-burner cooker hob, and as I circled the arrangement of slabs and cauldrons I thought how I would have loved to spend time getting used to cooking on this system – finding the hotspots, fuelling each fire, working with the draughts, understanding its subtleties.

On shelves around the central hearth were all manner of wooden vessels: bowls, kneading troughs, staved buckets and tubs, barrels, half-barrels, firkins and costrels. A large millstone, presumably hand operated, lay in a timber tray alongside a series of wooden pestles and mortars. The dairy was equally impressive, and obviously a large part of the farm enterprise. A vast selection of churns, plungers, presses, pats and moulds must have been a collection from numerous farms in the region. Each mould was engraved with detailed patterns, but these motifs were not just for decorative purposes. They were the means by which Icelandic dairymaids branded their butter for market, and as I examined the rich variety of symbols and letters it wasn't hard to envisage the familiarity and pride behind each design.

At the heart of the more modern kitchen (c.1890) was a large cast-iron cooking range, the centrepiece of a room that had obviously benefited from the Industrial Revolution. All the same processes were represented here as in the earlier kitchen, but hand-cranked mills and cream separators illustrated how far into everyday life mechanisation had permeated. I felt a pang of reassurance as I caught sight of a coffee bean grinder, a small comfort to my modern sensibility. But it played on my mind as to whether I really had the stomach to carve out an existence in such a harsh environment. There was certainly no lack of comfort within some of the timber-panelled rooms. Upstairs, the bedrooms were cosy and compact. Box beds, heavy blankets and eiderdown quilts looked like they would induce a peaceful night's sleep. Although the furniture downstairs was minimal it was well made and set within a very liveable space. In one of the reception rooms was a range of bureaus and chests, and even a piano. A wall-mounted clock with decorative panel inlay proudly displayed the date 1760. A place to read, write and play music. What more could one wish for?

5

THE SKEP-MAKING BEEKEEPER

MANKIND HAS ENJOYED a mutually beneficial relationship with the honeybee since at least the middle of the third millennium BC. This relationship has afforded us the priceless sweetness of honey, the medicinal qualities of propolis and, in beeswax, one of the most versatile domestic substances. In return, we have protected and propagated the honeybee. Left to their own devices, wild bees tend to favour hollows in tree trunks for their hives, but in the process of domestication human beings have adapted receptacles and designed purpose-built structures to house their tamed colonies. Our earliest evidence for bee husbandry, a series of bas-reliefs and wall paintings from ancient Egypt, suggests that these industrious little insects were kept in large clay cylinders laid on their side, with an entrance for the bees at one end and a removable door for human access at the other. In colder climates, archaeological evidence from north-west Saxony suggests that upright log hives, mimicking the bee's original home, were being used between the first and fourth centuries AD. From around the same period, a wickerwork hive and timber base, recovered from excavations conducted on the North Sea coast of Lower Saxony (modern-day northern Germany), is an early archaeological indication that people were using basketry to create homes for their colonies of bees. In Britain, documentary sources inform us that bees were being kept in wicker hives by the eighth century.

Known as skeps, basket hives can be made either of wickerwork (willow or hazel) or from straw stems twisted and bound together by a harder cane-like material. Little is known of straw skeps until the later medieval period when they feature frequently in manuscript illustrations, but there was a possible example dating to the twelfth century recovered from excavations undertaken at Coppergate in York in the late 1970s. The anaerobic conditions provided by submersion in water (it had been starved of oxygen) had preserved a series of twisted plant stems loosely bound together by a thin stem-like material, and from the same layer thousands of fragments of honeybees were also recovered. Unfortunately, we can't be sure about the exact relationship between the bee fragments and the twisted rope-like material, therefore its interpretation as a bee skep must remain conjectural. Yet some support for this early identification comes from the word itself. Derived from the Old English *sceppe*, which in turn is derived from Old Norse *skeppa*, it implies an etymological derivation from a period of English and Norse intermingling, making York – a city and hinterland where Anglo-Saxons and Vikings learned to live side by side from around the ninth century – a possible candidate for the origination of skep beekeeping.

A *sceppe* in Anglo-Norse England was a very specific quantity of grain, a volume that, like the Winchester 'bushel' in its early days, was probably reliant on a conventionally made vessel for measuring it out. Whether the first occurrence of bees taking up residence in baskets was a deliberate or accidental act, it would surely have been observed that they thrived in this environment. Why? Because straw has incredible properties of insulation and during the cold winter months of hibernation, as the colony huddles together for warmth, the higher temperatures achieved can result in less die-off of bees and therefore a greater number of workers to kick off the hive's work in spring. As any beekeeper will tell you, get a head start in spring and bumper crops of offspring and honey await you in the summer and late autumn. Until the mid-nineteenth century, straw skeps were the hive of choice for many beekeepers in Britain and other parts of Europe.

In the nineteenth century, against a climate of ever increasing scientific interest, naturalists set about the study of the honeybee with a view to better understanding the inner workings of the colony, undoubtedly with one eye on improving yields of honey. One of the problems with skep beekeeping was that you couldn't really see what was going on in the hive, and although as early as the eighteenth century some skeps were equipped with side windows to view activity, beekeepers were still largely reliant on watching the behaviour of the bees as they came and went from the hive entrance to ascertain the overall health of the colony. By the middle decades of the century, Lorenzo Langstroth in the US, Jan Dzierżon in Poland and Edward Bevan in Britain, among others, were exploring how the hive could be adapted to favour the scientific study of bees. The results of their endeavours were a series of timber box hives with moveable combs – some fixed in frames, some hanging from bars – all with the same principle: the hive could be opened from the side or the top and individual combs taken out for study and harvesting. Further adaptations included queen excluders, which restricted where in the hive the queen could lay her eggs, and supers, boxes of frames that could be placed on the top of the hive to capitalise on expansion in honeycomb as the colony grew. These types of hives are now ubiquitous, certainly in Europe and the US, and have encouraged beekeepers to develop an array of techniques that allow them to manipulate the bees, control their reproduction and maximise honey production in an unprecedented manner. Great news for beekeepers, but were they good for bees?

WHAT APPEALS MOST to me about straw bee skeps is the basic nature of the material required for their construction: straw and bramble. The tools needed are also of the simplest variety: a pocket knife, a small spar hook, a needle made from the leg bone of a goose or turkey, and a sleeve or funnel made from a length of cow horn. Anyone living in rural pre-Industrial Britain would with very little hardship have been able to find the tools, resources and time to sit down and make their own skep. Anyone, therefore, could theoretically have had a beehive in the back garden.

Today, if you stand out in a field of wheat it's likely that the plants wouldn't come up much above your knees. But a century ago farmers grew what are now called long straw varieties, so named because their stems could reach heights of up to six feet. Since the Second World War, biologists have worked towards creating a plant that redirects the growth hormones and nutrients from the stem to the ear, resulting in shorter plants but more numerous and heavier grains in the head. The new varieties better fitted the requirements of a time when there was less demand for straw and more demand for food. As a by-product of early cereal crops, long straw had a number of uses around the farmstead and in the world of craft – in effect, a free and widely available material which, when kept dry, was remarkably durable.

Bramble too is often more widely available than many farmers or gardeners care for, but it's not the sprawling thickets of bramble which spread inconveniently across precious grazing that are of use to the skep-maker. The bramble required for skep cane must be well managed in a location where it competes for light with taller shrubs, such as thorn or spindle, so that it grows a trunk of considerable height, which allows its flowers to break through the hedgerow canopy. It does this to attract the bees for pollination, and so that its fruit can lure birds, which, in devouring the sweet swollen blackberries, become the unwitting carriers of the bramble's seeds. I have cut and drawn bramble plants which, having spent their short life competing with mature hawthorns, have reached a length of thirty feet. But the skep-maker needs to be selective. If the stems are too thick the cane they produce can be too robust and, while this material might be useful for other jobs around the homestead, it's too inflexible for binding the straw in the skep.

Usually, a stem the thickness of a little finger is ideal, and it's better to cut it at around five feet in length. Anything longer than that becomes impossible to wield in the making process. It's also important that the stem is free from branches as these create knots in the wood, which in turn create problems in the splitting process required to convert the stem into three or four lengths of cane. An extremely sharp, thin-bladed penknife can be used for slicing the stem down its length and into quarter strips, but the more skilful and effortless technique of guiding the split with a sharp spar hook would be preferred. Once split, the pith inside is exposed and must be removed before the cane can be used. If left in it will rot over time and loosen the bind of the cane on the straw, making the whole skep unstable. In all of this, the most efficient method is to let the cut bramble stems lie submerged in fresh water for a good few weeks. The material as a whole becomes much more pliable, so that the splitting can almost be done with your fingers and the rotted pith scraped out with a spoon. A speedy way of splitting cane is to use a cleaver, a short wooden baton with its ends carved in such a way that once an initial split has been made with fingers or knife, the baton can be drawn through the stem to carry the split evenly to the end of the length. However, stems can only be split this way if the pith has been soaked and weakened to the point where it offers no resistance.

Letting nature do the hard work for you – by soaking – might massively cut down processing time but it does require prior planning and the appropriate type of water source. If you get the soaking wrong, you can ruin your whole cane harvest. If the water is too stagnant, for example, it can rot the cane wood and weaken it. Equally, if it gets wet and dry and wet again too often it can become brittle. There is some merit in working with freshly cut and processed cane, for as it seasons it shrinks and binds the skep tighter together. Whether you soak it or not, it's advisable to remove the barkwood. This can be done using a stropping technique, with the foot securing the cane to the floor as one hand holds the other end taut and the back of the knife is run down the inside of the cane from top to bottom in a slow and steady action. This, in effect, pushes the barkwood away from the heartwood. The most time-consuming aspect of skep-making is sourcing the cane. As with every job, skep-making is made all the more easy by working with the best materials, and a good-sized skep can require in the region of two hundred feet of cane.

THE MAKING OF a skep is an exercise in control and rhythm. At all points the hands must be working to constrain the materials, manipulating them with the end form in mind. For the beginner, photographic examples may be needed. But for the seasoned skep-maker, a subliminal aesthetic, a beauty dependent on how the skep is intended to function, guides the work. The straw must be lightly soaked so that it's pliable in the hands and forgiving in the twist. It must be dressed too, and it's best to spend a good twenty minutes, while the straw is still in the sheaf, combing the leaves off the stems to achieve the smoothest possible finish. Keeping the thickness of the straw layers even is another challenge and requires the aid of the cow horn to funnel the straw stems together. This action in particular must be conducted with a rhythmic regularity for the straw to flow seamlessly from the sheaf, through the funnel and into the bind. The cane must also be wetted so that it resists splitting and fraying as it's woven through the rolls of straw. The technique of stitching is known as lip work and involves passing the lip of the cane into the void created in the roll of twisted straw by the spearing of the straw with a leg-bone needle. The cane is then drawn through tightly and bound around the next roll, or course, of twisted straw before being secured to the previous course through the same lip work stitching technique. In this fashion, the skep spirals into life. Shape and form are the maker's choice, but are conceived as part of a slow realisation rather than an act of preconceived draughtsmanship.

ARE SKEPS MAKING a comeback? Many of the criticisms levelled at today's unsustainable agricultural systems could just as well be made at modern methods of beekeeping. During the last hundred years or so, we have developed a tendency to overexploit and ride roughshod over natural forces rather than working with them, and changes in beekeeping have been no different. There is no doubt that the change in hive types introduced in the nineteenth century significantly improved our understanding of the honeybee. We have been able to observe at close hand the developments within the comb, the laying cycle of the queen, the behaviour of worker and drone and the causes and consequences of disease. But these developments have also put us in a position where we can manipulate the bees in our pursuit of maximum yields. In some ways bees have become our prisoners, or slaves.

Although skeps precluded inspection and analysis on a scientific level, perhaps the biggest criticism made by modern beekeepers is that to get to the honey you have to destroy the brood in the comb. Frame hives, by contrast, have compartments and excluders that prevent the queen from laying eggs in certain parts of the comb, and it's from these frames that the modern beekeeper can harvest, keeping clear of the queen and the next generation of bees. With skeps, without removable parts and a means of keeping the queen away from some of the comb, it's much harder to harvest the honey without compromising the brood and the hive's means of reproducing itself the following year. It's not strictly true that the whole colony needs to be destroyed, for the outside edge of the skep can be carefully harvested in order not to disturb the nucleus combs. But this is a fiddly process, and it's undoubtedly more efficient and higher yielding to harvest all the comb inside, destroying the colony in the process. Modern beekeepers might well take issue with this, but would the beekeepers of old have been so concerned?

I don't think so. And there are a number of reasons why. First, the egg-laying honeybee queen has a working life of only four or five years. So, if you have a skep with a queen approaching retirement, you can confidently harvest for honey knowing that the chances of her having another productive year are slim. In essence, rather than let her continue for another year and find out too late that her egg supply has dried up, you cash in early and, in doing so, the cleaned-out skep will be ready and waiting for a new swarm the following year.

It is here that another supposed shortcoming of the skep actually works in its favour. The natural way bees reproduce is via a swarm. In the height of summer, when the days are long and hot and the pollen is in full flux, the queen will decide that she and around half the colony will flee the hive in search of a new home. She will leave behind a hive stocked with honey stores, a workforce of young bees and, most importantly, an egg that will shortly hatch and provide this new colony with their queen. Thus one colony becomes two. The problem for modern beekeepers is that, as a consequence of this halving of the workforce and the time it takes for both colonies to get back up to full strength, honey yields at the end of the season are significantly lower. So, if you can suppress the swarm, preventing the midsummer break in production, you stand to bring in a bumper yield of honey at the end of the season. Modern removable frame hives allow a range of manipulation techniques to be used by the beekeeper to prevent the queen from swarming. Skeps, on the other hand, do not. In fact, a contrary attitude is taken by the skep beekeeper where she is actively encouraging her bees to swarm, so that year on year she has a healthy supply of new bees to stock the skeps that she has harvested the honey and comb from the year before.

Skep-keeping relies, therefore, on an entirely different philosophy and economy of beekeeping – one almost in opposition to modern methods. Some people would argue that yields are far lower, but it's a much more apicentric – bee-friendly – way of producing honey, and I think that more traditional methods can find economic parity with the methods born out of nineteenth-century improvements. In modern hives the conscientious beekeeper, in the interests of the colony, will take little more than one-fifth of the honey surplus, leaving the rest for the bees to sustain themselves during the winter. Commercial beekeepers will take more, replacing the shortfall with a synthetic sugar-based solution, and then wonder why the colony is in ill health come the following spring. The skep beekeeper effectively achieves the same ratios when they sacrifice one in every five colonies because, as explained earlier, the queen is in her fifth year. Thus, one fifth of the honey surplus is taken for human consumption. There may be less overall because of swarming, but, in many ways, you want them to swarm – it's good for the bees.

There is another element to this comparison, though. A beekeeper using removable frames, employing mechanical centrifugal methods of extraction to draw the honey out while leaving the comb intact, will argue that the comb from the frames can be reused by the bees in the next year, saving them the job of building new comb, and thus allowing them to get straight on with honey production. Cutting and harvesting the comb from a skep destroys it completely, but the plus side is that you have more beeswax at your disposal; in medieval and early modern times, before the introduction of petroleum-based candle waxes, the beeswax was almost as precious as the honey.

There are other benefits in making the bees build fresh comb on a more regular basis. With fresh comb there is less chance of disease carrying over, and although modern-hive beekeepers will criticise skep beekeeping because it precludes access to inspect for diseases, it may be that in a world of predominantly skep-based beekeeping, there is less need to take such a cautious approach to disease. This is because one of the key benefits of skep beekeeping is the active promotion, rather than suppression, of breeding. We're constantly warned about the risks of disease to honeybees, and while removable frames allow closer and more regular inspection in order to take preventative steps (which usually amounts to pouring toxic chemical products over the frames), if we weren't advocating such intensive methods in the first place the threat of disease might not be so great. As with any livestock, as soon as methods become commercial the risk of disease rises. Perhaps most importantly, if we have, since the beginnings of the modern removable hive, suppressed swarming and reproduction on a regular basis, to what extent have we retarded the species' ability to evolve and build resistance to new strains of disease?

The objections to skep beekeeping are cast in terms of yields and disease but are mired in a mid-nineteenth-century pseudoscientific way of keeping. I think this obscures a not necessarily more intelligent way of beekeeping, but a different way. Basically, we dispensed with *cræft* in favour of science. And there are certain economics that come with more traditional methods. The market for beeswax is one. At the point when cheaper alternatives started to flood the market, it became more lucrative to focus the enterprise on the production of honey. Also, taking one-fifth of the surplus from one hive is very different from sacrificing one of five hives. With skep-keeping, you have to keep more hives on the go. This requires more space, but also more hives. One of the main selling points of a cedarwood box hive would have been its longevity. Treated with varnish or an oil-based paint, you could get a good forty years of life out of it. But they were initially expensive. Skeps, on the other hand, were inexpensive, made from what were then essentially free materials and in a time when labour was cheap. So, there are many variables that need to be considered, as well as a series of value judgements, before we decide on the best method of keeping bees.

THERE IS SOMETHING *cræfty* to skep-beekeeping, and I say this off the back of successfully keeping bees in skeps. I got into bees in the mid-2000s, and like many before me it all began with the discovery of a couple of old and dilapidated beehives. The ones I found were tucked away at the edge of a small wood on the high ground above our cottage. I put in an enquiry with the local farmer, whose land they were on, and found that the hives had been there for years, and that he'd lost touch with the beekeeper. Fair game, I thought, and after a couple of wheelbarrow trips, I shipped both hives down the hill and onto the workbench of my back shed. Over the course of a couple of days, the best bits of both were salvaged to make a fairly solid hive. Now all I needed was some bees.

They were not hard to come by. There were two enormous colonies within a mile or so of my house. One had taken up residence behind the shiplap panelling of an eighteenth-century granary building down the road, and every year it seemed to be going like the clappers. I was even more familiar with the other one – it lived in our chimney. Every year these colonies swarmed, and both were big events. I remember the first year I experienced the swarming of our chimney colony. I was sat in the house working at my desk when I heard what sounded like a helicopter landing on the roof. As tens of thousands of bees set off for a new life elsewhere, they hurtled around the chimneystack in a cloud of reverberation. This event happened, year on year, at almost exactly the same time – in late June – and it was this swarm that I intended to catch.

And I did. It involved a ladder precariously balanced on the very top of a thorn tree, some pretty vicious stings to my wrists, and a long wait to see if – in shaking the swarm from the branch and into my prepared cardboard box – I had actually captured the queen. But it was all worth it: two days later, my rehabilitated hive was busy with the comings and goings of worker bees, and my beekeeping days had begun in earnest. Over the following years, I acquired more hives, and at its peak my enterprise grew to eight colonies. Of course, more hives meant more inspections, more kit and very much more time to carry out the best practice advised in various beekeeping guides. The whole business started to get a little too serious and I found myself spending as much time administering chemicals and purchasing sterilising equipment as watching the behaviour of the bees around the hives. Meddling, tinkering, manipulating – it all became a bit too industrial for me, and as someone who has been gifted with a palate that lacks a sweet tooth, I wasn't even that interested in the honey.

I decided to let nature take its course for a few years. Discovering that the nearest domestic bees to mine were a good two and a half miles away, I wasn't too concerned about the spread of disease, but in any case, it was time to let the bees run their own affairs. During this period my colonies never dropped below five, and on a couple of occasions I had bumper crops of honey. I very much enjoyed laissez-faire beekeeping and chose to spend the time I would have spent interfering in their business otherwise engaged in the sowing and planting of bee-friendly flowers – especially clovers, sunflowers and lavenders. Along with my brother, I also started exploring alternative methods of beekeeping. While he set about building top-bar hives – a simple but effective hive for resource-challenged communities in Africa – my interest had been ignited by the find of a skep in the attic of an abandoned *fermette* in the Creuse region of central France. At the time, I was researching the old stone-built huts of the *bergers* – shepherds – whose traditional life rearing sheep on the hills of the Limousin region had been almost completely replaced by the industrial production of beef for the French fast-food industry. Taking refuge in the abandoned farm during a sudden downpour, this chance encounter with a traditional form of straw hive committed me to a new form of beekeeping. After weighing up the possibility of carrying this ancient artefact as hand luggage on my return flight home, I instead resigned myself to making my very own version back home in England.

In a fit of craft puritanism, I decided that my skep would be made entirely from materials grown within my own garden. This was going to be my *cræft* – to see if I could turn the natural products of a half-acre of English downland into honey. What had seemed like a wonderful and romantic idea in the beginning, over eighteen months later had turned into a labour of love. My skep became something of a major project. It would have been possible to build an extension on the side of the cottage in the time it took me to finish my skep. But this is largely because, in seeking to use exclusively home-grown materials, I'd opened a proverbial can of worms. In the first instance, I needed to grow the straw. I managed to obtain grains of the right long-straw variety without too many problems, but it's best grown as a winter wheat, a plant that is sown and germinates in late autumn, lies dormant over the winter months before growing on in the spring, flowering in the summer and going to seed the following autumn. Grown this way, the plant is stronger, the stem taller and the end product better suited to skep-making. So, by the time I harvested my wheat in late August a whole year had already passed. But I hadn't wasted my time. Over the summer months I spent any free moments I could experimenting with different types of bramble cane – all sourced from the hedgerows bounding our garden plot. As a consequence, I had a substantial amount of prepared split cane to bind together the straw. Finally, I could craft my skep. And it took about three hours to actually make.

But I still wasn't happy. Because skeps are made of straw, bound in a circular fashion, they're not particularly water resistant. While they can brave the odd shower, if left out permanently they can get soaked through, which would spell disaster for the bees inside. Skeps have always required some kind of shelter. Visit any old country house, look closely, and somewhere, cut into the walls of the grounds, the walled garden or an ancillary building, you will find what are called bee boles. These are purpose-built alcoves for housing skeps, providing the perfect waterproofing for the insulation properties of the straw. Lower down the social order is a range of techniques that can be used to achieve the same thing. On a trip to the Museum of Welsh Life in St Fagans, I came across an ingenious arrangement of large slate slabs used to create a series of sheltered shelves into which a good number of skeps could be housed. Working our way down the social order further, I've seen old photographs of skeps perched on logs in orchards with a variety of large flowerpots, old pans and off-cuts of tarpaulins providing shelter. I decided I'd go for a purpose-built bole, and using some reclaimed bricks I created a large free-standing alcove with one concrete paving slab as a base and another as a roof. After a week of rain, I inspected the interior of the skep and found it full of cobwebs. If other insects were flocking to it then it was certainly dry and warm enough for bees.

But every time I looked out of the kitchen window I winced at the look – the aesthetics – of my hive. I didn't mind the old bricks too much but the concrete slabs were just so ugly. They grated against the whole ethos of my skep-making, one that depended on only the most locally sourced materials. I needed to find an alternative. Stone is in such short supply on the chalk downlands that trying to make a natural stone covering was out of the question. I considered making a timber roof, but in the end resorted to the most traditional method. I still had a substantial quantity of wheat straw left, so I decided to make what is called a hackle – a thatched straw cone that sits on top of the skep. This took a while, but not as long as the thatched outer wicker frame I'd also decided to make as an extra insulating wall. In the end, it became the Rolls-Royce of skeps. In total, it had probably taken me the best part of two months of part-time work to complete. I was starting to feel like a true Arts and Crafts practitioner.

Eventually, I managed to get some bees into my hard-earned skep. In mid-February, I was asked to take the large colony from the granary building down the road. It was due for a refurbishment and, wishing to pursue a bee-friendly approach to their removal, the contractors had given me a call. It was a battle cutting the combs from the void behind the panels, delicately placing them in the skep, and trying not to upset the very drowsy bees. I knew it was touch and go as to whether they'd survive, and whether the queen would pull through such an upheaval. But trying to move them was better than having the pest-control officer destroy them with poisons. As April came, the hive started to show promising activity, and by late May they were exhibiting some of the vigour they'd shown back in their old granary home. The bees thrived in the skep, and continue to do so. My whole beekeeping enterprise has now been stripped down to three hives, and of the three I have absolutely no doubt that the bees in the skep hive fare the best. Without fail, they're the first colony to get going in early spring, they produce an adequate surplus – which I harvest by means of placing an extra skep on top of the brood skep – and they are entirely capable of looking after themselves.

The *cræft* in beekeeping is not in the meddling in the bees' affairs but in the preparation of their home.

6

TAMING THE WILDS

ORIGINALLY, WE WERE hunter-gatherers, or, as some scholars now prefer, gatherer-hunters, in recognition of the fact that we probably spent more time crouching on our hands and knees gathering up wind-fallen nuts and fruits than we did manfully belting through the undergrowth, spear in hand, in hot pursuit of quarry. But at some point, approximately twelve thousand years ago, we domesticated certain plants and animals and adopted a more sedentary lifestyle. Exactly why we did so remains one of the greatest questions of the human story, marking as it does a profound shift in our relationship with the natural world and triggering a series of monumental social changes. No longer would the vast herds of the animal migrations determine where we overwintered, nor would the bountiful harvests of the alluvial plains dictate where we set up camp in the summer. As we tamed the wilds and coerced the natural world to fit in with our designs, we gave up our nomadic lifestyle and gradually became farmers. Then, as we produced surpluses and began to covet land, so our populations boomed and the arena was set for the human display of hierarchy, power and conflict.

Dubbed the Neolithic Revolution by the archaeologists of the 1920s, this period of domestication and settlement was by modern standards anything but sudden. Some believe that the adoption of farming practices across prehistoric Europe took as many as two millennia, and in Britain the so-called Agricultural Revolution of the Neolithic (the New Stone Age) is accepted as roughly covering the period from 5,000–4,500 BC to 3,000–2,500 BC. And the story of how humans domesticated the landscape around them doesn't end with the late Bronze Age agriculturalists. The varied processes by which we cultivated wastelands or wildernesses to grow food or graze livestock continued developing right up until the mid-twentieth century. While the most fertile and workable land in Britain had already been cleared of its primeval temperate rainforest by the end of the prehistoric period, the more advanced agricultural practices of the Romans, driven by a need to support the emergent cities of the *civitates* – a burgeoning state infrastructure and a pan-European trade in goods – extended the agrarian topography of the British landscape. The subsequent collapse of the Roman Empire very likely saw many fields and farmsteads return to wilderness, with the fertility of centuries, if not millennia, lying dormant until the later part of the first millennium AD, when agrarian developments were once again on the rise. Throughout the medieval period landscapes of extremes found themselves the victim of humankind's tenacious ability to manipulate the natural world. Marshes were drained, wetlands reclaimed and the tradition of assarting – the clearing of trees and shrubs to create farmland – further ate into the ever dwindling moors, marshes and woodlands of the British Isles.

This is a story that can be told through the main tools of domestication and the means by which it is sustained: the walls, hedgerows and ditches that make up the skein of complexity in the British landscape we see today. These boundaries were for centuries the seasonal concern of an army of agricultural labourers whose *cræft* it was to ditch, wall and hedge – practices that went hand in hand with the agricultural traditions they facilitated. To control the land, to subdue its tendency to return to the wild, it needed to be stock-proofed in order that crops could be protected, but also so that the gluttonous attentions of herded cattle and sheep could be concentrated on particular areas, grazing and manuring them in preparation for future arable cycles. Today, wire fencing has become the stock-proofer's barrier of choice; its immediacy, and requirement for little annual maintenance, has seen its widespread adoption regardless of landscape type. Everywhere you look today you find pig wire, chicken wire, barbed wire and supporting wooden stakes. On an aesthetic level, the character of landscapes once defined by their stock-proofing barriers is steadily being eroded. The drystone walls of the Peak District, the hedgebanks of the Devon fieldscape and the sweeping hedgerows of the chalkland arable are all giving way to this homogeneous – and lifeless – means of demarcating space.

The damage of wire fencing is not just aesthetic. It is also historic. Wire fencing, once the breakthrough technology of the industrial agriculture of the nineteenth century, very quickly came to undermine traditional British farming. Its global export meant that for the first time vast stretches of prairie in the Americas and outback in Australia could now be enclosed and exploited at speed, and with little investment of money and manpower. The result was that livestock could now be farmed on an unprecedented scale, which in turn meant that British farmers, saddled with their historic hedges, ancient walls and medieval ditches, simply couldn't compete. And like all modern quick-fix systems it was effectual, at least in landscape terms, only in the short term. While wire fences are quick to put up and require little maintenance for a couple of decades, they have a finite lifespan and, over a longer timeframe, require imported materials to retain their structural integrity. As the collapsed posts and twisted wire observed on any countryside ramble will attest, these posts might not be replaced any time soon. In short, the continuing wire-fencing boom of the twentieth century is distinctly un-*cræft*-like.

The opposite might be said of walls, hedges and ditches. The initial investment in the creation of such barriers is substantial – especially when undertaken by hand. They need maintenance: a hedge needs laying every six to seven years; a drystone wall needs occasional restacking in places; and the base of a ditch needs redigging. But they are monuments that last and, most importantly, they are born from the very earth they partition. A hedge needs no imported materials, since year on year it grows its own. A drystone wall needs only restacking. A ditch needs nothing more than a spade to refashion its profile. They have proved faithful servants to British farmers for centuries, if not millennia.

Remarkable as it may seem, there are stretches of drystone wall on the Pennines and on Dartmoor that have been archaeologically proven to date from the earliest phases of agricultural development in the Bronze Age. A method of dating a hedgerow, called Hooper's Law after the scientist Max Hooper, involves counting the different plant species it contains along a given stretch. With every new species over and above an initial three varieties taken to represent a hundred years of its existence, in most areas of Britain it has been demonstrated that many of our most substantial hedgerows date back to the medieval period. Ditches providing the main axes of coaxial field systems in the Cheshunt area of Hertfordshire have also been dated to the late Iron Age, on the basis of pottery recovered from their primary fills. Alongside this longevity, hedgerows, walls and ditches contribute other vital services to the farm as a whole. They are a crucial source of biodiversity in an otherwise potentially monocropped landscape; they provide shelter from wind, rain and sun; and ditches play an important dual role in the management of irrigation.

WE TEND TO think of hedgerows as permanent fixtures, essentially static manifestations in the landscape. But to operate effectively, they have to be nurtured on a cycle of six to ten years. This involves laying, pleaching or plashing, a process in which the new growth of the last six or so years is folded down into the line of the hedge and woven into other laid branches or between regularly placed stakes. Left to its own devices, a hedgerow species, like any shrub or tree, will seek to grow to its maximum height. In becoming more treelike, the shrub develops a canopy of leaves that prevents sunlight from reaching the emerging shoots of lesser shrubs, stunting their growth. The result is substantial gaps around the base of the larger shrubs, and if exploited by animals these gaps can widen at an alarming rate. Neglected for too long, a hedgerow can become nothing more than a row of irregularly spaced spindly trees, and their shade depriving the hedgerow floor of other shrubs to plug the gaps.

So what does pleaching a hedge involve? The answer is an extremely sharp billhook (like a hooked hatchet), a hedger's mitten, a bow saw and a considerable degree of craftsmanship. If the hedge is in reasonably good shape and has received proper management in the first place, the various species will have developed in such a way that a plentiful number of shoots are growing from the base. In the first instance, you don't really want the thickness of the shoots you intend to lay to have grown beyond the width of your lower arm – if they have, it's because they haven't been pleached recently enough. These shoots need to be removed altogether. For the remaining shoots, the trick is to make a cut with the billhook at the base of the shoot, far enough through so that the trunk can be easily bent down without snapping it clean off – usually between a third and halfway through. Depending on the style of hedging – and each part of the country has its own methods – this shoot is bent down to anywhere between parallel or diagonal to the hedgerow floor, effectively barring passage. It's important that the cut doesn't go so far through the shoot that too little of the barkwood is left intact. The barkwood is the means by which the shoot will continue to grow, transporting the nutrients and growth hormones from the roots to the growing tips which, now that the shoot is laid flat, will emerge along its length, providing a stock-proof barrier over the coming years.

Certain rules need to be observed, such as always laying hedges uphill. As sap rises, you need to ensure that the places on the laid branch where you want new growth to develop are always above the top of the roots. Also, keep your cuts clean (hence the sharpness of the billhook). That way the shrub stands a good chance of healing quickly and getting on with the job of growing. One swift and accurate chopping motion should be enough to achieve the required forty-five-degree cut. Hacking away with a blunt tool to generate the right depth and angle of cut can result in a messy wound to the branch, which is more likely to trap water and rot the shrub at its base. Another tip is to match your forty-five degree cut with another cut, at right angles to it, to remove an angular chock, or spur, and smooth the line from the laid shoot, which again helps to shed water.

Every now and again you need to remove branches altogether, and this is where a certain level of confidence is required, along with the ability to see four or five cuts ahead. When I first started hedging, most of my day would consist of long pauses as I procrastinated over whether or not a particular branch should be removed. As time passed, it became more of an intuitive process, a simple and logical deduction of which branch would stay and which would go in order to create the best hedge. There are various reasons why you remove branches altogether. If a branch is too large it can, when laid, dwarf other laid species below, blocking out the sunlight and suppressing growth. Evenness and uniformity are crucial too. Certain species of shrub grow at variable rates, as indeed do members of the same species. After ten years, that variability may need bringing into line.

Also, you might want a better-placed but smaller shoot from the same shrub to develop as the main shoot. Leaving in its larger sibling will mean that it's always battling for its share of growth hormones. Remove the larger one and the smaller one becomes the lone recipient of a disproportionately large root bole and thus grows much more vigorously.

Of course, this is the kind of knowledge you gain when you've worked in an industry long enough to observe and learn from the fruits of your labour. In the ten years or so that I've been dabbling with hedging I can now revisit certain stretches and see what worked and what didn't. And a lot of my work could have been improved on. But to the regular hedger, the learning curve would have been steeper and the mastery of the craft achieved sooner. The ultimate goal is a hedgerow that is so well laid and developed that it becomes almost a solid wall of foliage. Such hedges require a different form of maintenance – one more akin to the regular pruning of a garden hedge. On a large scale, however, pruning shears would make heavy work of agricultural hedgerow management. A freer and more fluent technique known as brushing or combing must be used to get through the miles and miles of hedge needing treatment.

Over the last ten years this has been something I do at least twice a year with the larger hedgerows around the garden. For this job you need a hedging hook. Like so many vernacular tool types, the part of the country you come from influences the shape of the tool. I've experimented with several different types, ranging from the large short-handled crescent-shaped hook to the long-handled shallow-hooked variety. It very much depends on the types of hedgerow species you're working with, which are largely dictated by the soil, which is itself dictated by the underlying bedrock, hence the regional variation in tool types. I've settled for something in between – a sort of short-handled scythe-shaped hook – which when sharpened slices admirably through the outer foliage of the hedge, tightening up the whole barrier. At first, I had to work to the shape of the existing hedge, one very much the creation of the chattering hedge trimmer – square, ordered, boxy. But over time, my technique has moulded the hedge into a smoother and more rounded figure. It's a technique I usually sum up as a series of three motions. First, to cut the part of the hedge that grows from the ground to waist height, I adopt a swiping motion somewhere between a tennis player's backhand and a cricketer's straight driving shot. To cut the section between waist and head height, a tennis player's forehand with heavy topspin is required, and to cut the part of the hedge level or above head height, it is very definitely a cricketer's hook shot.

In the bible on the subject, *Hedges for Farm and Garden* (1950), J. L. Beddall states, 'The hedger has always been a craftsman and takes his place in the economy of the countryside. In the days of the self-supporting and self-sufficing manors his was as much a craft as the smith's, carpenter's and carter's.' Today, it's easy to overlook that skill. Hedges have become more of a nuisance to farmers, landowners and councils, and the standard approach has been to machine flail, indiscriminately, and to such a degree that many are sick from canker and neglect. Our relationship with hedges has come to reflect, in the words of the writer and environmentalist Roger Deakin in *Wildwood: A Journey Through Trees* (2008), our 'disdain for nature'. Such sentiments are so drastically far removed from the views of the sixteenth-century farmer and poet Thomas Tusser, who, in his *Five Hundred Points of Good Husbandry* (1573), points out many of the positive aspects of hedgerows that we have so woefully chosen to forget, including the wealth they bring to the land through the facilitation of crop rotations, and that in themselves they 'hath plentie of fewell and fruit'. Furthermore, they provide shade and shelter for crops and animals, not just from the cold and drying winds but from the scorching sun too. As John Worlidge observed in *Systema Agriculturæ* of 1668, livestock without shade 'lost more flesh in one hot day than they gained in three cool ones'. All this is to say nothing of the environmental benefits. Their biodiversity supports a huge array of species – birds, mammals and plants – without which the countryside would be a poorer place. And this is where the true *cræft* lies: they are so *fitting* within a landscape. They facilitate a degree of intensification without compromising the health and well-being of their immediate environment.

FOR MOST OF my life I've lived within landscapes where hedgerows – or at least, what's left of them – have predominated as the means of controlling stock and partitioning natural resources. This is largely because, on the chalk downlands of southern England and the Weald clays of Sussex and Kent, there is little else by way of materials. In many parts of the country, however, there is no shortage of an altogether more permanent type of resource: stone. Look at any geology map of Britain and you will be stunned by the variety of geological strata that make up our island's bedrock. And it's these different stone types, where they are hard enough to be used for building, that make up the human character of the landscapes they have been quarried from. It is their colour, their grain, their shape and form in the construction of walls and buildings that make each area of Britain so identifiable – a regional character that we lose with the employment of mass-produced or imported brick, slab, slate, concrete and barbed wire.

So-called because they're erected without a bonding mortar, drystone walls are made from stone alone and must stand exposed to the worst that the elements and livestock can throw at them. As such, it's the craft of the waller that binds them together, giving them the structural integrity they need, and it's through the waller's skill that the properties of a local stone are observed so nakedly in all their glory. No two geological regions of Britain are the same, and the nature of the stone very much dictates the nature of the wall, along with the techniques used in its making. The hardest stones are the granites from places such as Dartmoor and Bodmin in the south-west of England and various outcrops throughout the Cambrian and Grampian mountain ranges. These are the most difficult to fashion and one needs to adopt a 'less is more' approach in trying to smack them into usable shape – you'll sooner bust your wrist than form the perfect bedding stone. Then there are the gritstones such as the famous Millstone Grit from the Peak District of northern England. These stones will fracture along a plane, but they're still clunky and determined in their nature.

In North Wales, the nature of the ragstone is such that it splits so beautifully you can make slates as thin as 5.5 millimetres. A strong and versatile material, slate was exported all over the country during the nineteenth century as a superior roofing material, but in the process of production only one tenth of the quarried material ended up in the final product. Left lying around, the surplus waste was soon incorporated into the walls of farm and farmstead, facilitating the growth of the North Wales sheep industry. Of all British stone, Carboniferous Limestone – particularly that of the lower Oolitic strata – a band of stone that runs north from Portland Bill in Dorset, through the Cotswolds and up into the north-east of England – represents arguably the best stone to work with. It splits well, and in both directions, hence agreeable rectilinear blocks can be achieved with little effort. But you won't ever hear a waller bemoan his local stone. It would be like denying your own birthright.

The craft of drystone walling doesn't begin the moment you start smacking the stone. There is an enormous amount of preparation work, beginning with some serious consideration. A drystone wall is a huge undertaking, and you must really have a need for one. When your mind is made up, first, choose the line you want your wall to take through the landscape. It's always a good idea to make best use of topography: read the landscape and observe the contours and terrain. When stock-proofing, for example, it's better to site the wall at the top of a break of slope rather than the bottom, so that sheep don't use the bank of slope as a springboard to jump the wall. A good waller is also conscious of the effect a wall will have on the landscape, and, in anticipation of the natural inclination of soil to creep down the valley side, will bolster, deepen, buttress and thicken the wall in anticipation of accumulation.

Ground clearance goes hand in hand with wall construction. Very often a new wall will be prescribing an intake of rough ground, land intended to be brought under the plough and worked down to support crops or grazing. Many of the existing stones will need to be cleared and any boulders too big to be handled will require breaking up. The easiest way to do this is to light a localised fire under the central part of the boulder (so you need to dig a channel below it) and subject it to intense heat. Once you see a change in the colour of the stone closest to the fire, hastily pour a bucket of cold water onto it. The stress of expansion to sudden contraction in only a small but central part of the stone will create thermo-fractures throughout the boulder. I've witnessed heated boulders the size of crouching men reduced to a pile of rubble with only one strike of a heavy sledgehammer. It's an incredibly satisfying feeling and by far the least labour-intensive method, but you can't control the line of the fractures. To turn a boulder into a series of face-stones – straight-sided stones that give the wall a flat face – the method of boring and wedging is required, creating a fracture by design. It's more laborious, but it can save time at the point of construction. It's often the case that in the clearance of boulders from a field you don't want to reduce them too much in size. Ideally, as soon as the stones can be manhandled they can then be rolled into the line of the intended wall and used as foundation or base stones.

For the bulk of the wall, stone must be brought in from the native hills. Quarried stone is invariably better than the rich pickings of scree found at the base of mountain cliffs and crags, or the Dartmoor clitter, the fractured boulder stone from the tor. Buried deep under the earth's surface, quarried stone has been protected from the sun, rain and frost of millennia and as such retains a youthful resilience. But most importantly, try and ensure that your source of stone is uphill from your work. It's easier to bring mountain rock down than to haul river-borne boulder deposits up. And at one tonne of stone needed for every square yard, planning your lines of supply becomes a vital part of the whole process. It's what's known as the 'economy of handling', and in many respects my childhood years poring over piles of Lego blocks gave me a small advantage when I first began drystone walling.

Back then, I would set apart the necessary blocks for certain specialist parts of the job, weigh up the available material by eye and think through the possibilities before embarking on the build. This is a crucial part of a stone-waller's *cræft*. There's a limit to how much stone can be quarried and delivered to site – you certainly don't want to be taking any away at the end of the job – so the material must be scoured before work begins. Good face-stones, trig-stones, foundation stones, cornices and tie-stones must all be set aside and distributed evenly along the course of the intended wall in order to achieve uniformity of appearance and make best use of the available material. 'Painting and decorating is nine-tenths preparation,' my father sagely informed me as I embarked on my very first home makeover project. The same rule applies to drystone walling.

Then comes the digging of the foundation. Your trench should be deep enough to come down onto relatively solid ground – if not bedrock then a heavily compacted subsoil. Not only does the foundation provide the necessary structural footing for the rest of the wall it also acts as a guard against burrowing animals, which over time can undermine the wall and cause it to collapse. Tools are also an important consideration. Today, we need to think a lot less critically about which tools to use: we just load up the 4x4 with as many as possible, including the highly specialised. But back in the old days, when the majority of wallers walked to their place of work, with or without a barrow, a much more judicial approach was adopted. They worked with tried and tested classics: a waller's hammer, a pick, a shovel and an iron bar (to use as a lever).

Many wallers will set up frames in the form of a template at either end of the stretch of wall they're working on. Between each frame string lines are levelled to indicate relative heights between the courses of stone. At first, I thought this method a cheat for achieving the most aesthetically pleasing wall. Surely, all craftsmen should work by hand and by eye and need not rely on guides and levels? The hand–eye mantra may work for crafted objects that can be picked up and turned in the hand, where the shape and form can be appreciated up close. But to do the same for a drystone wall you need to step back a considerable distance to take in the whole edifice and check for any glaring errors. This takes time that can ill be afforded when you have an entire mountainside to enclose. As such, the template frame saves time.

I remember working on a section of wall one afternoon when the stones just seemed to slot into place. It was only at the end of the day, when I stepped back to marvel at the lengthy stretch of wall I'd completed that I realised I'd followed a line bending out from the desired direction of the wall: a cumulative error derived from each stone being only slightly out of line from the previous. The wall might well have functioned adequately but it looked wonky. It had to go. Ultimately, if you're out of sight over the brow of a hill, battling wind and rain to get a section done, appearance probably matters less. But if you're building a stretch across the front of the lord of the manor's croquet lawn, you'll probably want to get your template frames out.

Drystone walls comprise two outer-facing skins and a central core. Face-stones are laid in rows called courses. In each course, as in brickwork, each stone must 'break the bonds' of the stones in the previous course: one stone must overlap two and two overlap one. It sounds quite straightforward but can get a little tricky when you factor in a third requirement: the overlap with both the internal stones of the wall core and the stones on the opposing face. Stones that face the wall on both sides as well as passing through the wall itself lend structural integrity to the whole and are known in different parts of the country as through-stones, tie-stones or jumpers. These are arguably the most crucial, and any waller worth his salt will spot a tie-stone a mile off and set it aside for that purpose.

If all this wasn't complicated enough, there are other design features that the waller needs to adhere to. For example, ideally you need to work to a 'batter', a receding slope where the base of the wall is slightly wider than the top. Walls built in this style are more likely to have structural integrity, avoiding the risks of higher stones toppling. But with a lengthy wall, a batter can save on material: tie-stones which are too short for tying the wall at its midpoint can be used at the top, and from a sheep's eye view the perspective makes the wall seem just that little bit taller. If at all achievable, there should be a good run-off of rainwater too, because if any moisture running into the centre of the wall gets trapped and then freezes it can fracture the surrounding stones and over time cause internal structural problems.

It's as important as the requisite overlap between stones that the outward faces of the walls are as smooth as possible. This is not about appearance. Animals love to scratch, and as a result any protruding point or edge will quickly become a place to rub against. If a two-hundred-pound ewe really gets the itch she can dislodge the protruding stone and destabilise an entire section of wall. If protruding edges are a feature of your walling style, and the whole flock gets the itch, then you might as well kiss your wall goodbye. You need to work the faces of the stones down so that they're flush with the line of the wall. This is where the waller's hammer comes in handy. On one end is a square hammer used for more substantial breaking. You might, for example, have a fairly large stone which is not long enough to be used as a tie-stone, and equally doesn't have a flat face on either of its sides. Using the hammerhead, you can smack the stone forcefully in a line along the centre of the stone with the ambition of breaking it clean in half. Successfully achieved, you will have two good face-stones. The chisel on the other end of the hammer is used to refine the faces so that they're even more flat.

Some stones will need corners smacking off in order to sit comfortably in the wall, others will need striking with their grain to make them thinner so that they sit more agreeably within the height of a course. All this results in a series of angular chocks and other discarded fragments. These are not to be wasted. Some can be used for what one Dartmoor waller referred to as trig-stones. Not every stone sits comfortably on the one below it and will often rock ever so slightly. A trig-stone can be used as a wedge to chock the stone and prevent unwanted movement. The remaining discarded material makes up the core of the wall and is known as the hearting, which indicates how important this part of the whole structure is. Good overlap between the stones of the heart and the face is as important as it is anywhere else in the wall.

A stone that is easier to work can result in a style of walling with many decorative but nonetheless functional features. Cornice stones, which slightly overhang at the top of the wall, can serve as drip-stones, carrying the worst of the water away from the face-stones. And at points of access, all manner of gates and stiles can be fashioned to allow passage through the wall for humans without compromising on stock-proofing efficiency. These require stones of very particular shape, such as quoins (cornerstones) and scuncheons – jambs for a gate to slap against. Obviously, finishing off the top of the wall is one of the most important parts of the job, and this task is made much easier with a stone that's more amenable to splitting. Purposely fashioned coping-stones, copestones, capstones or toppers can be made up – heavy slabs with a rough pitch worked onto each side. Or smaller stones can be set on end in what is sometimes called a cow-and-calf configuration, alternatively small and big to give an almost crenellated effect.

Much of what I've described here are the techniques and forms that are afforded only when working in perfect conditions: good stone, easy availability of material, ideal landscape and wealthy backers. For this setting you need look no further than the Cotswolds, and in the short time I've spent playing with Cotswold stone, as opposed to Dartmoor granite or Welsh ragstone, it really is child's play. The stone is worked so easily you can almost ask it to break where you want it to. But in many ways, this means you have to up your game to get anywhere near the standard of wallers who have hammered bluestone all their lives. But if *cræft* is also about being on time and to budget, you have to ensure that you make the stones do as much of the work as possible. Selection, therefore, is as important as the skill of hitting the stone. As one waller said to me, 'You can tell a bad waller by the size of the pile of gravel at his feet' – he has clearly spent too long hitting the stone into shape.

A local drystone waller doesn't get to choose his stone – the differences to him are entirely academic. You may well find yourself less ambitious when using stone that is harder to work, in countryside where there is less wealth in the land – not just in terms of fertility but in terms of capital and access to markets. But walls still need to function and, at the extreme end of the scale, a particularly talented waller is required for what at first glance appear to be the most amateurish of walls in the Hebrides. Comprising a single-skin thickness and made from essentially unworked rounded boulders piled loosely on top of each other, these walls appear to be peppered with holes and have a slight wobble to the touch. But these are intelligent walls. They're relatively quick to build but you need skill and sharp eyes to work with these hard granite stones. The gaps are purposeful: the strong Hebridean winds blow through them rather than blowing them over. The lack of stability is also by design: it plays to the sheep's innate fear of the wall toppling and therefore encourages them to stay away. The trick is to create a deliberately unsteady wall with just enough wobble to dissuade the sheep, but not so much that the wind blows it over. It's a fine line, and one that only the finest *cræft* practitioner can walk.

Drystone walling at first seems to be a straightforward stone-on-top-of-stone process: a labour more than a craft. But what is undoubtedly clear is that many factors lie behind the successful execution of a drystone wall that will stand the test of time. These range from the simple economy of handling through to the ineffable ability to reshape stones and lay them in the direction against which they meet a matching stress from neighbouring stones: to make them immoveable without any recourse to bonding agents such as lime mortar or cement. But the real *cræft* of the drystone waller – and the hedger, for that matter – is not a sleight of hand, a single moment of brilliance. This *cræft* is a sustained assault on a landscape, a refashioning, as slowly and methodically as the land itself changes through natural processes. Patience is essential. You have to wait not days, not weeks, but years to see a stretch of wall erected, and decades to see a field system emerge.

I N ANGLO-SAXON ENGLAND during the tenth century, when the king granted an estate to one of his subjects a charter would be drafted – ink on vellum – to record the transaction. The charter would detail, in Latin, the nature of the grant, the location of the estate, the benefactor, the beneficiary, and the nature of the dues owing on the land as a result of the grant. At the bottom of the page would be written a witness list, comprising the names of all those present at the granting of the estate. And, finally, appended to the base of the manuscript would be a boundary clause: a list of boundary marks, written in Old English, defining the outside edge of the estate. These landmarks are colourful, characterful and often graphic in their descriptions of the Anglo-Saxon landscape, and, arguably, it's not until the Romantic poets of the eighteenth century that we again see the English landscape described in such rich and vivid detail. Many of the more permanent landmarks, such as hills and rivers, can still be found in the landscape of today. But others, such as the 'old willow' or the 'fox-hole', have been lost to us for ever. Some landmarks have terms so obscure that they remain either indecipherable or the subject of speculation and debate. One such term is *wyrtwala*.

Translated literally, it means 'root wall'. It crops up in boundary clauses in charters referring to estates throughout the southern counties of England on varying geologies, where it is found in woods, heaths or on arable land. Whatever this *wyrtwala* is, it isn't fussy about the terrain it covers. It has a sister term too, *wyrttruma*, meaning 'root firm' or 'root strong', and this term is as ubiquitous across a similar area. Place-name scholars, Anglo-Saxonists and landscape archaeologists are pretty sure that some kind of linear boundary feature is being referred to. Granted, 'wall' is a bit of a giveaway, but equally, when the term is used in boundary clauses, the boundary itself is often described as running 'along' the 'root wall' or 'root firmness'. But what exactly does this linear structure, made fast by roots, consist of?

As with many Anglo-Saxon phrases, it's often the case that you need to read the term literally – say what you see. When out walking I often stumble across stretches of hedgerow planted on banks of earth where the species are so old that their root mass is exposed through the bank, and as the earth erodes away over time more of the roots are exposed to create, essentially, a root wall. Today, in modern arboricultural parlance, this process is known as adaptive root morphology. The examples I've come across are likely to be two or three hundred years old, based on Hooper's Law (the maturity of the shrubs) and the fact that many of the boundaries within which they sit are already there on mid-nineteenth-century maps. Evidently, these were hedgebanks that by now had reached full maturity. It's tantalising to think that the Anglo-Saxons of the tenth century, in their choice of terms, were describing hedgebanks initiated by their forebears some two or three hundred years earlier, barriers that now retain their strength through the reinforcing properties of the roots. They represent a time-honoured, entirely natural way of dividing up the landscape, and one with a vision that stretched far beyond their maker's lifetime.

Wire fencing has become symptomatic of the new and

uncrafted relationship we have with our agrarian landscapes. In our desire for quick capital gains we have come to rely on short bursts, hitting our landscapes hard in smash-and-grab bids to turn minimum investments into maximum profits. The cost has been our traditional commitment to the land, a long-term attitude in which the maintenance of drystone walls, hedges and ditches was a recurrent feature. In racing so quickly to the quick fix of wire and posts as our method for taming the wilds we have opted not to continue to invest skill and knowledge in more traditional means. We have grown economically accustomed to not setting aside the finances to dedicate manpower to a means of organically and entirely sustainably managing the landscape, to say nothing of the environmental benefits that hedges and stone walls have for the biodiversity of their environment. For all its quaint beauty, the British landscape was no match for the extensive and industrial farming systems that wire fencing permitted in the new world. But our walls, hedgerows and ditches have created a lasting, indelible and distinctly regional infrastructure which enabled us to set about taming and constraining the wilds for a thousand years and beyond – with a need for little more than a waller's hammer, a hedger's hook and a ditcher's spade.

7

WEFT AND WARP

AS I WRITE, the archaeological excavations of a Bronze Age settlement recovered at Must Farm in Cambridgeshire are drawing to a close. In many ways, the hard work of interpretation is only just beginning as the archaeologists set about analysing, dating and conserving the many finds that were recovered from this remarkable excavation. First discovered in 1999, the site was subjected to archaeological sampling in 2004 and then full excavation in 2006, and since then it has captured the imagination of amateur and professional archaeologists from across the world. The reasons for this global interest lie in the level of preservation of organic remains, as well as the way the settlement came to an end. The site consisted of a series of round houses constructed on a wooden platform supported by a number of timber piles driven into the wetland soil below. This watery realm, which provided both an element of security and access to crucial natural resources for its inhabitants, also served as the means whereby all the archaeological materials at the site came to be preserved. As a fire took hold in the buildings and drove the inhabitants out, the platform collapsed into the shallows below, resulting in the various organic materials (as well as inorganics, such as pottery and metal) being permanently waterlogged, starved of oxygen and surviving in a state of suspended decay.

This sudden catastrophe, the nature of the abandonment and the subsequent preservation has seen the media dub the Must Farm site 'Britain's Pompeii', enthralling many from outside the world of archaeology. From within the discipline, however, Must Farm has enabled a number of experts to answer critical questions about everyday life in the Bronze Age, told through the domestic artefacts, and also the level of their technological aptitude. What is emerging is the story of a people who had an incredible understanding of the natural world around them and who were capable of an extraordinary level of technological sophistication. Among the everyday finds – such as buckets, beads, plates, troughs, wooden boxes, handles, storage jars, cups and bowls, some still containing the residues of food – is evidence of building styles, tool types, weapons and wheels.

The two that captured my imagination are an extremely fine collection of textiles and a panel of interwoven hazelwood known today as a hurdle. At around three thousand years old, the textiles represent the best, largest and earliest assemblage we have in Britain. During the fire, the fibres of the cloth were carbonised which, having then been smothered in fine, non-porous sediments and waterlogged to this day, allowed for remarkable preservation. The recovered fragments comprised plant material, chief among this was flax, which is still used to make fine linens. But there was also evidence of how it was processed. Hanks of the original plant material before it had been combed out were found alongside balls of processed thread and remnants of weave and twine. Overall, these textiles were of an excellent standard, with some of the finest threads having the thickness of coarse human hair, which allowed the Bronze Age weaver to produce rather exceptional cloth. The hurdle was of interest simply because I've made these things myself, and in the process have come to see the importance of this craft to farming and food production across the world, right up to the early twentieth century.

But both these artefacts are also the result of a particular form of craft: a shared technique that can be summed up in the Anglo-Saxon words *wefta* and *wearp*. First appearing in the written word in the early eighth century, the practice of weaving a weft of material in a horizontal fashion through a vertically set warp of the same material is at least as old as the evidence recovered at Must Farm. Strictly speaking, these are terms that are applied to the weaving of textiles, but the principles of wefts and warps can also be seen in the hurdle where the horizontally set rods (wands) are woven through upright rods (sails or zales).

IT'S AMAZING TO think how weaving, an industry so global in the modern age, had once been so local – a basic domestic craft practised in everyone's front rooms. Many of the processes of carding, dying, spinning and weaving were carried out in the same place – perhaps even by the same person. Evidence for domestic cloth production can be found from every single period of British history right back to the Neolithic, and in some remote parts continued well into the early twentieth century. Certainly, in rural communities it was a useful sideline during the quiet times of the agricultural year, when the days were short and long evenings could be spent spinning on the wheel and casting on the loom. But dedicated weavers producing textiles for commercial purposes were probably around from at least the medieval period. With the growth of towns from the tenth to the thirteenth century onwards, the woollen cloth industry played a large part in the economic success of Britain's farming communities, from the small hill farmers of the mountain regions to the wealthy monastic houses of the lowlands.

Weaving was one of the first crafts to industrialise properly. It was run on factory principles with a clear division of labour and adopting early forms of mechanisation, particularly with the introduction of the power loom in the late eighteenth century. Certain processes in the preparation of yarn and finishing of cloth could be adapted to milling, with power provided by running water. Early larger-scale enterprises were found close to medieval market towns, capitalising on existing watercourses, supplies of local wool and passing traffic. Wales, for example, with its access to good wool and good water, benefited in this period. But the Industrial Revolution, and in particular the shift to steam power in the mid to late nineteenth century, meant that even larger enterprises set up on new sites where access to the sources of fuel, a willing workforce, building materials and proximity to the main lines of industrial communications (canals and rail networks) were more of a concern. Marginal workshops still carried on, but the full globalisation of the industry in the twentieth century was to have a substantial impact on the regional British cloth industry.

In less than a fifty-year period, from 1926 to 1973, the number of Welsh wool mills fell from 250 to 27. Two world wars didn't help either. The end of the First World War dealt the first sucker punch. As a government surplus of flannel and blankets flooded the market, sales of these domestic staples never really recovered. In some ways this event is a testament to the longevity of the crafted object itself: make something well, and from good materials, and it will last beyond one generation. But what do you do when everyone already has a highly durable blanket? This is every weaver's dilemma, though it rarely affects the standard of their work. Not compromising on quality is a philosophy of production that is, of course, the antithesis of the modern textile industry whose capital returns rely on heavy turnover, a throwaway culture in which every item has a planned obsolescence and barely lasts a lifetime, let alone long enough to be passed down to future generations. I don't think we have altogether lost the tradition of passing down garments, blankets or valued fabrics. I have my maternal grandmother's curtains, for instance; my sister has a beautiful 1930s dress that belonged to our father's mother; and my brother has a lovely Welsh woollen blanket that once adorned the parcel shelf of our maternal grandfather's old Rover P6. But because we no longer appreciate the huge amount of effort and work that once went into making cloth traditionally we don't see the inherent value of a handcrafted weave. We don't understand the *cræft* knowledge that underpins it.

I'm not sure that the people of the Must Farm site would have rued not rescuing their fine linen as they watched their stilted homestead burn to the ground. They too might have inherited the cloth from their parents or have intended to pass it down to their descendants. Certainly, there is physical evidence from the medieval period – some 2,300 years later – that highly prized tapestries, vestry and liturgical textiles were passed down from generation to generation of clergy. When historical sources emerge in the medieval period, in the form of inventories and wills, we gain a much clearer insight of this practice at all societal levels. One document, which I came across when researching land tenure in tenth-century Wessex, provides a fascinating window into the conveyance of important and obviously cherished garments and cloths. Commonly known as Wynflæd's Will, the document survives as an eleventh-century copy of an original probably written at the mid-point of the tenth century. In its surviving form it is a remarkable text and a rare opportunity to examine the property, chattels and household staff of a wealthy Anglo-Saxon woman.

Wynflæd makes many gifts to relatives and churches throughout the kingdom of Wessex, within which she owned a number of dispersed estates. She bequeaths land, cash sums, valuable jewellery and tableware, as well as freeing slaves and donating livestock. It is, however, the bequests of cloth and clothing that are most interesting. Perhaps the most prized and important textiles Wynflæd gifted were her offering-cloths, on which donations to the church were presented. But she also grants various members of her family and household a set of bedclothing, two chests containing her best bed-curtain, a linen covering and all the bedclothing that went with it, her best tunic, the 'better' of her cloaks, her black tunics, her best holy veil and best headband, a clothes chest, a little spinning box and a slave described as a 'woman-weaver and seamstress called Eadgifu'. Land and cash are the usual subject matter of an Anglo-Saxon will, and in this respect Wynflæd's is no different from that of any of her male counterparts. But the listing of these most practical of items, many of which come at the very end of the will – items that an old and probably very wise lady felt necessary to include – is a touching indication of how cherished and valuable they were. It's an indication of the expense of time and effort that went into producing hand-woven fabrics in the age before the spinning wheel and the power loom.

WYNFLÆD'S COLLECTION WOULD have almost certainly been woven from raw materials produced locally, most likely on her own estates. The wool would probably have come from the sheep herded on the downlands throughout her many manors, while the linen will likely have been produced from the flax plant, which we know from documentary and archaeological evidence was grown in the Anglo-Saxon period. When you think of the damage a linen tea towel takes in the process of drying dishes and being washed at high temperatures, it's hard to imagine that this material is derived from a plant. The flax plant, *Linum usitatissimum*, is grown in much the same way as a cereal plant, with long thin stems supporting a head that flowers a lovely pale blue and produces, if it's allowed to mature, linseed from which oil is pressed. But if you harvest flax in much the same way as hay, before the seeds ripen in the head, you will trap a good deal of nourishment in the stem, providing the strength from which cloth is produced. It's the 'bast' fibres from the inner barkwood of the plant that are spun to make fine linen thread. Various processes, such as 'retting' (soaking), 'scotching' (thrashing) and combing, are required in its production, and the end product is gloriously smooth and incredibly durable.

There is sound archaeological evidence that flax was grown in prehistory to produce textiles and, although awaiting the full publication of the analysis of the Must Farm textiles, it seems likely that they will turn out to have been woven from the flax plant. It's staggering to think that today we still use plant fibres, as we did over four thousand years ago, to make materials for both luxury and everyday use. Ireland, arguably, produces some of the finest linen in the world. The climate, environment and a rich heritage of cottage and local industries are largely responsible for this tradition, but a huge debt is also owed to the Huguenot immigrants of the later seventeenth and early eighteenth centuries who improved standards of growing, spinning and weaving.

And then there is wool. For me, the single best illustration of our early national dependence on wool can be seen in the diverse range of sheep breeds that emerged in the late eighteenth century. For such a small island we enjoy remarkable regional variations in climate, soil and bedrock, and these undoubtedly had an effect on the development of sheep breeds in different areas, together with the social, cultural and economic impacts of the developing wool market from the early medieval period onwards. And different breeds produce different kinds of cloth. For example, varieties like the Devon Longwool or the Leicester Longwool, hailing from bleak and wet parts of the country, tend to produce a fleece with a long 'staple' (the length of the fibres) that is lustrous but inelastic. Needless to say, it makes a cloth that works well in bleak and wet parts of the country. In drier and warmer conditions, short-wool sheep – such as the downland breeds of Sussex, Hampshire and Dorset – thrive and produce a fleece with a short, fine and elastic staple. Again, this weaves into a finer cloth for a warmer and drier environment. As expected, the mountain breeds of Wales and Scotland produce the coarsest fleeces, which are usually used in the making of tougher woollen products such as trench coats and the famously durable cloth known as serge.

This isn't always the rule, but it does mean that British wool producers and cloth makers were in a position to match their products to a wide variety of terrains across Europe as markets opened up in the thirteenth and fourteenth centuries. And we shouldn't forget that on a single fleece there can be as many as ten grades of wool, ranging from the long staple of the back – used to make finer fabrics – to the coarse 'britches' around the legs, which would be collected and sent off to the rug maker. So a range of products could be produced on a very local scale to match differing needs from the local breed of one's own region.

IT'S DIFFICULT TO know where to start in the process of converting raw fleece and flax into cloth. For wool, the production line begins with selecting the right ram to put to the flock. The ewes must be in good condition when they receive him. They must birth and feed their lambs well and save enough energy to produce a fine fleece. Setting aside the part played by the shepherd, wisdom is required in all the many steps needed to make a fine textile from raw fleece. As the saying goes, to become a master weaver you have to have 'been through the mill', that is, to have seen the entire process from start to finish. You have to understand how fleeces translate into fibres, how fibres translate into yarns, how both respond to dyes, and then assign the appropriate use of these materials to a garment with a particular function.

First, the fleece would need to be cleaned, and by far the best way to do this was to walk your sheep to the nearest fast-flowing stream and get them to take a bath. It sounds like more work than necessary; after all, why not shear first and then tub-wash by hand in a place of your own convenience? The problem is that you need *a lot* of water. More to the point, a fleece dries a damn sight quicker when it's being carried around on the body of a ewe grazing in the early summer sun. Like most animals, sheep are reluctant to do anything other than drink at the water's edge. So you need to use a fair bit of coercion to get them to take the plunge. At the end of the day, there's no substitute for wrestling them into the deep, jumping in yourself and agitating the fleece to release the dust and grit.

I did this once, back in my mid-twenties, and it was a glorious experience. On the edge of the farm was a small stream, which turned a sharp corner at the bottom of the meadow and had on its inner bend the perfect beach to launch the unsuspecting flock from. It also had a natural depth on its outer bend to submerge them in. I'm convinced, after the initial shock of the cold water, that the sheep enjoyed the experience as much as I did. They went almost entirely limp as I took them to a depth where they could just about touch the stream bed. Being relatively buoyant, they required only a supporting hand under the muzzle as the other hand worked its way around their body, ruffling and combing. A dense cloud of fine dust flowed out of their bodies, and I continued with each ewe until the fleeces ran clean. Hauling them out was no mean feat, and they staggered off under the weight of their saturated fleeces into the plush meadow to graze and dry.

Once sheared, further washing can be undertaken to condition the fleece. This was traditionally done by soaking it in lye, an alkaline concoction made by passing water through wood ash. Solutions of diluted urine – or a mixture of urine and lye – would also help lift the grease and shift the dirt like a natural soap. Some caution is required at this stage, because if you wanted the end product – say, an overcoat or a Guernsey or Gansey sweater – to have a degree of water resistance, you wouldn't want to lift out too many of the natural oils. Today, of course, these greases are entirely lifted out and then, if required, added back in at the end of the process. But leaving too much oil in the fleece makes it too viscous to separate out into its component fibres. This process is known as carding. The modern machinery today comprises rollers surfaced with very fine steel combs, which rake the wool to produce a continuous light and fluffy mass of uniform raw wool to be mechanically conveyed to the next process.

Before mechanisation this was a tedious job that needed to be done by hand. There is evidence from medieval manuscript illustrations that carding would have been carried out using carding boards – very fine comb-like brushes that were drawn back and forth across each other in a technique that teased out the individual fibres. It's open to debate whether teasels, a thistle-like plant, were used for carding. Certainly, well into the nineteenth century, when areas of south-west England were commercially growing teasel plants specifically for the wool industry, we know that they were used to raise a nap on the finished cloth, to fluff up the fibres and make the fabric softer and more luxurious. But whether or not the natural tines on the seed heads of the teasel plant were strong enough to card stiff and often quite oily wool remains hypothetical.

However, during a visit to the National Wool Museum in Drefach Felindre in Wales I came across a 'teasel gig', a mechanically powered drum into which panelled compartments of teasel heads had been framed. The finished cloth was fed through a series of rollers and passed over the teasel drum, and it amazed me to think that up until the invention of fine steel brushes the best thing to use for raising the nap on finished cloth was the seed head of a cultivated weed. In a single evolutionary step this contraption represents a remarkable overlapping of technologies from our ancient and medieval past into an industrial modernity.

Spinning, where the carded fibres of the wool are twisted together to make a yarn, is probably the most iconic of the processes of craft wool production. It can't have been an easy job for Neolithic cloth makers. On excavated sites of this period, small fired-clay weights, exhibiting a hole in their tops to fasten a short stick, are thought to represent a form of spindle whorl. To the top of the stick is attached a bunch of woollen fibres. As the clay weight at the bottom is spun by hand, continuously in one direction, more fibres are let out and the spindle is allowed to slowly drop to the ground as a length of yarn is produced. Once the spindle hits the ground the yarn is wrapped tight around the stick and the process repeated until the stick is too swollen with yarn to take any more. This is then cast off onto another stick before another length is begun. We can infer that this was the method because these types of drop spindles have been used by many cultures all over the world, up until relatively recently.

My good friend and co-presenter in the many *Farm* series, Ruth Goodman, showed me this spinning technique one winter's evening years ago. It's the kind of thing you get up to in a remote cottage with no TV or internet access. I'd come back late in the afternoon from a long walk during which I'd collected pocket-loads of sheep's wool snagged on the barbed-wire fences that separated arable fields from an expanse of unenclosed common land. For want of a modern drop spindle, Ruth demonstrated rather admirably how the same results could be achieved with a heavy-headed wooden spoon. I spent every evening that week learning to juggle the spinning of the spoon and the feeding of fibres into the twisting yarn until my supply ran dry. After all this effort, I had a ball of incredibly irregular yarn barely the size of my fist, and as I pondered my pitiful attempt I felt myself desperately yearning for the invention of the spinning wheel.

There is evidence of spinning wheels in Europe as early as the twelfth century, and its widespread adoption throughout England by the fifteenth century was likely a consequence of the prominent part that wool played in the growth of the late medieval economy. These very early wheels look considerably different to the more modern varieties, and it's difficult to ascertain from relatively crude manuscript illuminations exactly what impact they might have had on the speed of the process and the quality of the finished yarn. Clearly, much of the powering and feeding of fibres was still conducted and manipulated by hand, but the large lightweight wheel banded to a much smaller bobbin – a spindle around which the thread is wound – effectively geared up the rotations, creating a greater and more consistent speed of spin.

A later technological development was the introduction of a 'flier', a small cradle-like device that spun around the bobbin at a greater speed. Spinning wheels with fliers had two grooves cut into the larger wheel. One groove carried the band that powered the rotation of the bobbin, the other the band that powered the rotation of the flier. The flier was fixed to a smaller pulley than that of the bobbin and thus spun at a much faster rate. As the fibres were fed through the flier they twisted as they wound around the bobbin, giving the yarn that extra strength and finesse.

Finally, comes the treadle. This really is an important development. By means of a crank action on the wheel, a footplate was pressed up and down with the feet to give power to the whole process. The spinner's hands, having been relieved of the obligation to power the flywheel, could now concentrate on the feeding of fibres, making spinning commercially viable at the level of the household. This was to have a radical effect on cloth production throughout Europe. The estimates are that between eight and twelve times as much wool could be spun with a wheel than with a drop spindle, and this increase in serviceable yarn enabled the weaving industry to blossom.

At this point, you would be forgiven for feeling fatigued at the sheer volume of work required to convert the raw fleece into yarn – and we haven't yet threaded a weft through a warp. But like so many crafts, good product is based on great materials, in this case all quietly and modestly being worked, incrementally increasing in quality and appeal until the final skeins of yarn are delivered to the weaver's hut. Here, it is the warp that is the true foundation of a good weave. Get this wrong from the outset and as the weaving progresses the fabric will very quickly lose shape. Tension is everything, and this is where the loom comes in, the means by which one set of threads – the warp threads – are secured in position in order that the weft threads can be interwoven to produce the weave. All this has to be done under tension because the weft and warp must be forced together and bonded on the frame.

Way back into prehistory we can only speculate what form of loom structures were in existence at the local level, but the standard of the Must Farm textiles suggests a pretty impressive bit of equipment. Tension is vital for a weave to work, and on later frames a windlass would crank tight a roller onto which the warp threads had been secured. But what I like about earlier looms is their reliance on the one natural force for tension that is a constant – gravity. To *cræftily* make use of this natural force, the loom must be set in a vertical position and the warps tensioned by a series of heavy weights. What we know of these warp-weighted looms is mostly derived from archaeological evidence, and although the recovery of the frame structure is exceedingly rare, the loom weights frequently survive by virtue of the fact that they were made from fired clay. Essentially, instead of securing the warp threads to the frame, or a hand-cranked roller, they are secured to rows of what look like ceramic doughnuts. These have turned up in excavations of the medieval period and right back, in Europe at least, to the Neolithic. Initial reports from the Must Farm excavations indicate the recovery of loom weights, and because of the nature of the preservation of organic material, it may be that some loom timber has survived.

Where looms have collapsed and rotted *in situ*, they tend to drop their weights in such a way that we can diagnostically calculate the number of weights, the number of warps, and thus width and thickness of cloth. A really good example of this comes from the excavations conducted at West Stow in Suffolk in the mid-1960s. Here, archaeologists exposed the best part of an entire early Anglo-Saxon village, uncovering a complex arrangement of halls, huts, ditches, fence-lines, hollows and pits. From Hut 15, a style of building known as a *grubenhaus* ('pit house') because the floor is sunken into the ground, rows of loom weights overlay a collapsed planked floor, which suggests that the whole complex had fallen into disuse and ruin. For every large, timber-framed hall recovered in the excavations, there appeared to have been a number of ancillary *grubenhäuser*. That Hut 15 seemed entirely dedicated to weaving might be an indication of early craft specialisation and commercial production.

Various reconstructions of the West Stow warp-weighted looms (and other excavated examples) have been attempted, and there is no doubt that this straightforward arrangement was effective. The weft threads were pushed upwards onto the warp threads in a manner that saw work progress down the frame towards the weights at the bottom. When more length was needed, the warp threads that had been wrapped through the loops of the weights could be let out. I've seen a demonstration of weaving on a reconstruction of a Viking Age warp-weighted loom and progress was slow to say the least. This was probably because these looms worked most effectively with two operatives, and in this demonstration it was just one lovely old lady from Denmark who was undertaking the work. But I made some rough calculations in my head, based on a number of assumptions concerning the size of a woven Viking longship sail. My conclusion was that the weaving alone, if conducted on a single loom, would have taken the best part of a year.

We're all probably more familiar with the appearance of the four-shaft handloom or floor loom on which the warp is set horizontally. Certain principles remain the same as those applied in the vertical variety: a mechanism needs to be in place so that every odd-numbered warp thread can be separated from every even-numbered thread to create an opening between the two groups known as a 'shed', through which a shuttle containing the weft thread is cast. The mechanism then needs to allow the alternate warp threads to switch position. This traps the weft thread and creates a counter-shed through which the returning weft thread is cast. On the warp-weighted loom it's safe to assume that this is achieved via a series of rods that rest on arms protruding from the loom frame. These can be pulled forward and set backwards, alternately, to create the shed and counter-shed. On the floor loom a series of treadles is used to raise and lower harnesses of an elaborate configuration consisting of shafts containing 'heddles' – frames through which the warp threads are passed.

In principle, one heddle contains the odd-numbered threads and another the even-numbered, and as they are moved apart, above and below each other, the shed and counter-shed are created. After each casting of the weft shuttle a baton device is used to compress the weft onto the warps, to ensure compaction and strength, and as the weave grows the length of the cloth can be extended beyond the length of the loom by letting out more of the warp threads which will, in advance of weaving, have been wrapped around a tensioned roller. You probably need to see these things in action to understand just how simple and yet effective they are. But be wary. If you ever visit a mechanical weaver's factory floor, take ear defenders. The racket is deafening. When I visited the National Wool Museum in Wales, it was with the express intention of hanging out in the weaving shed and soaking up the vibe. I lasted about ten minutes before I thought my eardrums were going to burst.

MY APPRECIATION OF wool was properly realised when we moved to a dilapidated house in the country with only a couple of old wood-fired log burners to keep us warm in the winter. The house was intended as a rented stopgap while we found our feet and recalibrated after a hectic London existence. We moved in during late July and I distinctly remember my first day there. It had been a roasting drive down from London in an old van loaded with our meagre belongings, and I took a shower as soon as everything was unloaded. We opened all the windows to air the place and a warm breeze blew through the house refreshing my cleansed body and invigorated spirit. This was a new beginning, I thought: yes, we were poor, but we were happy. What could possibly go wrong?

But as the wind changed direction in late November and a cold easterly buffeted the house, we battled to keep the temperature bearable. By early January, we'd done everything we could to improve the situation. I'd fixed draughty windows, serviced the wood burners, chopped more and better firewood and fitted draught excluders to the external doors. But there was one commodity that really made the difference, an age-old material that turned our chilly house into a cottage so snug that we ended up living there for over ten years. The answer lay in wool.

First, we backed up on declining the offer from my in-laws of some old-fashioned woollen rugs. These were spread liberally around each room to add snugness and to prevent rising draughts. We then purchased the thickest woollen drapes we could lay our hands on, and accepted my grandmother's kind offer of her lined woollen curtains. Knitwear became less of a fashion accessory and more a seasonal imperative. I found myself considering cardigans (for the first time since the Nirvana craze of the mid-1990s), tank tops and Fair Isle sweaters. I procured knitted jumpers from Ecuador and Iceland, flannel shirts from Wales and woollen socks from Italy. We scoured the local charity shops for woollen blankets and issued them to every bed and sofa in the house.

As our lives changed, I became more and more obsessed with woollen products and developed a wardrobe of particular favourites. Chief among them was an ex-British Navy issue boiled wool pea jacket and a Dunn & Co tweed suit. The coat I took everywhere. It was rough and ready, water-resistant, durable and incredibly warm. To this day it remains my go-to overgarment. I'd had the tweed suit for a number of years before moving to the countryside and worn it on only a few occasions. At the time I'd needed a cheap suit, and fast. I was a student in London, broke and in desperate need of a job. I'd looked in all the high-street chains for off-the-peg outfits, but all I found were expensive, cheap-looking two-piece suits cut in trendy cloth. I needed the reverse: something that was cheap but looked expensive. Then came some sound advice from a friend, a mature student who worked part-time cutting hair for the great and good of Sloane Square. 'Have you tried the Kings Road charity shops?' he asked. 'Designer gear, barely worn, and thrown out on a whim.' The next day, I raced to Chelsea and within half an hour had my hands on a beautifully fitted tweed suit for the bargain sum of twenty pounds. I was made up.

I'm anything but posh, and conscious of how in London a tweed suit can give a certain impression. But in rural south Wiltshire, among the ancient chalk downlands, it very definitely fitted. What amazes me about this suit is its adaptability to any situation. Whether hacking through the brambles, driving pheasants from the rough, networking at a local function or nipping up to 'town' (home-counties speak for London) for a business meeting, this suit always seemed to fit the bill. And unlike other suits I've bought since, it never seemed to lose shape. So I came to my present collection of tweed suits not through a desire to acquire a certain image but out of a search for the best, most long-lasting and most authentic material I could find. The material chose me rather than the other way round. I'm not necessarily obsessed with authenticity in craft production, but it's inevitably the case that the more original the process and the more locally sourced the materials, the better the product. And in my opinion there is no cloth finer, in Britain at least, than Harris Tweed.

If weaving from craft to industry has been affected by the tidal waves of the global economy washing out from the financial and mercantile centres of Britain, then these impacts were almost certainly felt least in the Outer Hebridean islands of Harris, Lewis, Uist and Barra. Here, a deeply ingrained resilience has served communities well and caused their traditional method of weaving to keep going, to ride the varying fortunes so many other forms of weaving businesses have fallen foul of. To be fair, the islanders have received their fair share of support from government grants to sustain and support business initiatives. But even so, the fame of their cloth has been hard earned and justly deserved. Anyone familiar with Harris Tweed will recognise the iconic Orb mark it is stamped with. This is what tells you that the product has been woven from wool sourced only within Scotland. This is wool that has been dyed, spun and finished on the Outer Hebrides and hand-woven, in their own homes, by the islanders of Harris, Lewis, Uist and Barra.

Already well established in the mid-eighteenth century, by the 1890s it was gaining a reputation in London as an extra-special cloth and a prerequisite for men of class and fashion. There's a small irony in the fact that the very people who benefited most from the Highland clearances, the wealthy game-shooting parties of Britain's elite and landed classes, were the trendsetters that put in motion the rising popularity of this fine vernacular cloth. Nonetheless, the Harris Tweed Association was formed by 1909 and the Orb trademark was registered. But it was hard to keep pace with the demands of this wider interest, as well as to compete with other weaving industries that were fast modernising. In a bid to revitalise and set the industry on a sure footing, semi-automatic, pedal-powered Hattersley looms – a bespoke design for cottage and home use – were introduced in the 1920s. By the mid-1960s, production peaked at over seven million yards of single-width cloth per year, with over 70 per cent exported, chiefly to the US. But in the 1970s fashions changed. The age of the true synthetic cloth was born as polyester and acrylic fibres rose to prominence. Mass manufacture and global sweatshops then caused costs to be driven mercilessly down, and throughout the 1980s the staid and outdated look of this cottage-industry cloth became anathema to the big-money glitz of Thatcher's Britain: too old-fashioned for the yuppies, too posh for the New Romantics. Between 1970 and the early 2000s, the number of Harris Tweed weavers plummeted from over two thousand to just under two hundred. An industry on which the islands had depended for nearly two centuries was left hanging by a thread.

And then something counter-intuitive happened. The very mass-manufacturing industrial model that had outcompeted Harris Tweed's method of production came to its rescue in the form of the American sportswear giant Nike. The company needed to differentiate itself from the other sportswear manufacturers during the training-shoe boom of the 1990s and 2000s. In seeking something authentic and original, Nike must then have somehow stumbled on Harris Tweed. They liked what they saw and placed an order that was to save the industry. The commission landed on the doorstep of Donald John Mackay, a master weaver who, working flat-out twelve hours a day, six days a week, could hit a target of around a hundred metres a week. Nike ordered 9,500 metres. In a show of true island togetherness, Mackay duly passed the work around to the other weaving concerns of the islands – and Harris Tweed as a global brand was born.

I don't know the exact thinking behind the Nike designers' decision to go with Harris Tweed. The story received a good deal of publicity in Britain, but would it have resonated with the popular press of Africa or Asia? Maybe not. I can only suppose that it was down to the quality of the cloth, a quality derived from strict adherence to a set of rules that qualify it for the Orb stamp: fine Scottish fleece, dyed, spun and woven by the hands of a select few in a tradition dating back hundreds of years. Even the look of the fabric is one born of the earth – the very soil of the Islands – with dyes taken from lichen, the tips of heather plants, peat soot, water lily, willow leaves and bog myrtle, all producing soft shades of browns, greys, purples, yellows and greens. They give the cloth a warm and wonderful appearance. But, as one islander said to me on a visit to Tarbert, Harris, 'The hand, and not the eye, is the best means by which to distinguish the superior quality of a handmade cloth from the mechanically produced variety.'

WHAT AMAZES ME most about Nike was their attraction to a material derived from such an ancient means of making. We've seen from Must Farm that the simple relationship between a weft and a warp thread is at least four thousand years old, but it is the application of this idea, this interweaving of materials in other similarly timeless crafts that interests me. John Seymour, in *The Forgotten Arts*, makes the candid observation that weaving likely had its origins in basket making. The theory runs that societies worked with ever finer materials until they realised they could produce something flexible, strong and capable of insulating the body. I'll go one step further, though. If weaving has its origins in basketry, then basketry has its origins in wattle-hurdle making.

Today, the wattle hurdle is the upmarket means to provide screening or fencing in a garden. It's considered the 'period' method for concealing unsightly but necessary objects, ranging from heating-oil containers to concrete outhouses or plastic wheelie bins. I'm not knocking the rural aesthetic of those who go for this approach to giving their garden an olde-worlde feel. After all, this support is preventing one of the most deserving of crafts from dying out altogether. Wattlework – the weaving of horizontal rods, wands or runners through upright stakes, staves, shores or zales – is a truly ancient and versatile technology that in Britain alone has found function from the very beginnings of farming in the fifth millennium BC. In the course of this journey it has served as a technique to contain livestock, a means by which properties are divided in occupational settings, and as the structural panels onto which daub – a mixture of clay, straw and dung – was smeared to create insulated external and internal walls in timber-framed buildings.

But it's the wattle hurdle, the stand-alone, moveable panel of wattlework that really intrigues me. Most commonly, it's associated with agricultural developments in the eighteenth and nineteenth centuries and seen as a classic rural craft of this period. There were substantial changes in the configuration of the landscape at this time in order to accommodate more productive systems of growing. In simplified terms, the traditional medieval three-field system of cropping was replaced with what is termed the four-course rotation. The additional crop in this system was the turnip, a vegetable that was able to fit seamlessly into the growing cycle for a number of reasons.

First, its deep taproot allowed it to draw on minerals and nutrients buried deep in the subsoil and therefore didn't compete with the other crops in the rotation, such as cereals and vetches, which drew their nourishment from hearty topsoil. Second, it provided a crucial break in the life cycle of pests and fungi that can affect cereal crops if they're grown on the same patch of land year on year. Finally, and perhaps most importantly, the turnip was not grown for human consumption but for the feeding of sheep in the winter months, when grazing was hard to find and a diet of dry hay alone could compromise the overall condition of the ewes. It wasn't just the benefits this crop had to the sheep and the next generation of lambs developing in their wombs. Perhaps the most significant contribution a crop of turnips could make to a farm was the levels of dung the sheep would produce while grazing on the crop in the field. In areas where soil was particularly thin and in need of enrichment, especially as land-improving farmers pushed the bounds of their arable ambitions to evermore marginal areas, the best method to add 'heart' was to have animals manure it directly from their own backsides. Crops grown on this land in subsequent years would produce significantly higher yields as a consequence of this free-range manuring.

So where does the wattle hurdle come in to all this? Well, there was a catch. If you just allowed your flock of sheep to roam at will across a ten-acre field of turnips they would graze it (and therefore manure it) in an erratic fashion. They would, of course, eventually finish off the whole field, but sheep don't spend the whole day grazing. They also spend a lot of time, like us, lazing around and sleeping, and for this they will always, whether for shelter or safety, tend to retire to a favourite part of the field. Thus, a large percentage of their dung would be concentrated in only one area. So, in order that they might obligingly spread their good stuff more evenly across the field, they would need folding (penning) to constrain their movement. In some parts of Britain, such as the Pennines or Peak District, where stone was in no short supply, drystone walls could be used for this task. Equally, hedgebanks were used in the small field systems of farming landscapes like those in the south-west. But the problem with these forms of partitioning is that they're permanent fixtures which, if introduced to lowland farms, would obstruct the ploughing, sowing, weeding and harvesting of the main cash crops grown in the other three years of the rotation. Better to have a form of partitioning that was removable and reusable.

This, then, is the historical and craft context for the wattle hurdle, an innovation that was integral to a system of farming associated with an age of improvement in the eighteenth and nineteenth centuries. This system of moveable pens aided the improved fertilising of the soil, vastly increased yields of cereal crops and, as a consequence, paved the way for the Industrial Revolution. So the story goes.

What happens, though, in the 1970s, when evidence for wattle panels was recovered from waterlogged deposits on the Somerset Levels? Dated to between 2,000 and 2,400 BC, these distinct panels were recovered overlying bundles of brushwood in a manner that suggested their use as a trackway across boggy ground. One, measuring around nine and a half feet by four feet, was lifted for preservation and under close examination was proven to contain diagnostic elements that demonstrated it was intended as a stand-alone panel. The results of the Must Farm excavations are eagerly anticipated, and it may well be that the panel of wattlework recovered there in a remarkable state of preservation could have features that indicate it was moveable. The question that therefore arises is whether we can legitimately back-project aspects of the agricultural advances of the eighteenth and nineteenth centuries into the prehistoric period. I'm not suggesting for one moment that the famed four-course rotation of the eighteenth century Agricultural Revolution has its origins in the Mesolithic, but it's not unreasonable to presume that prehistoric societies saw the value of controlled dunging and that they had the technology at hand to make it happen.

I REMEMBER ONCE ASKING a hurdle maker who they considered to be the best in the business. The curt reply was that a hurdle maker's hurdles are only as good as their coppice-work. The message was clear: fail to manage your coppice wood and you very quickly run short of good material to work with. So the craft of hurdle making technically begins with the planting out and year-on-year management of coppice woodlands. The term coppice has its origins in the Old English word *copped*, meaning 'topped' or 'polled', as in having the top cut off. Coppices differ from other types of woodland in that they're deliberately planted with species of tree that, once cut, will reshoot in a number of other locations on the trunk (stool). While willow, alder, ash, oak and chestnut will all do this, perhaps the most prolific coppice tree is hazel. Planted out, it will need to be allowed to establish for a good ten years before it can be cut for timber close to the base. A good clean cut will prevent any water ingress, and hence future rotting of the heartwood, and in place of the severed limb a number of other shoots will develop that, given a few years to grow, will be ready to cut for wattle-hurdle making. Often coppice woods will be planted out between standards such as beech or oak, tall trees whose summer canopy shades just enough sun to encourage the hazel to reach up towards the light. The shoots grow into straight and true rods, which are all the better to make hurdles with.

Hazel stools will take repeated cutting on a six- to seven-year cycle, and as a consequence of this pruning will live for an extraordinary number of years. I've heard tales of stools thought to be over a thousand years old, and although I can quite believe it as I've seen some enormous ones myself, I'm not certain that such claims can be verified through tree-ring dating simply because the heartwood of the original tree rots out as the regrowth creeps further outwards. As the demand for coppice wood grew in the nineteenth century, substantial coppices were planted out across many lowland areas, and particularly on the chalk downlands of southern and central England. I remember, during a summer spent wandering the ridges and combes of the South Downs, stumbling across a recently deserted hurdle-maker's camp in woods to the south-east of Winchester. Mature beech trees provided a dense canopy that gave the feeling of being in a giant medieval hall, echoes of distant woodpeckers rang between the trees and shafts of sunlight danced on the thin trail of smoke spiralling up from a residual fire pit. I was instantly fascinated by what I found, and with an archaeological brain started interpreting the arrangement of space, debris and unused rods to identify the work pattern of the hurdle maker.

Skid marks through an orange carpet of crisp beech leaves indicated that the raw material was being hauled onto the part of the site closest to the track. From here it was sorted into lengths and thicknesses and these groupings were then stacked against a kind of gallows structure – two uprights on the crutches of which a horizontal timber had been secured. It seems likely that at this point certain rods were selected out for splitting. A single short length of rod where an errant knot had caused the split to run out was the only evidence to suggest this. Unlike me, this hurdle maker made few mistakes at this crucial part of the preparation process. The next stage involved trimming each length to size, and this was evidently achieved using the stump of an ash tree, eight inches in diameter, and cut at waist height as a chopping block. I suspected that the location of the camp had been in part influenced by the suitability of this ash stump. There was a trampled path between stump and gallows as the hurdle maker walked between the two, and the pile of unwanted tops strewn around the stump to the right informed me that the maker was right-handed.

Next came the mould, which was positioned so that the worker's back benefited from the warmth emanating from the fire. The mould consisted of a trunk of timber originally some twelve inches thick, hewn in half lengthways, laid on its flat side, with nine holes drilled into its surface. The holes prescribed the shallowest of arcs, along a distance of six feet and the length of the intended hurdle. These were designed to hold fast the uprights (zales) of the hurdle as it was woven. As in weaving, the structural integrity of these warps was crucial to the overall standard of the hurdle. Small angular chocks lay scattered at each end of the mould, indicating that the horizontal rods, once woven through the zales, had been trimmed to perfection, flush with the ends of the hurdle.

At first glance, a wattle hurdle might appear to be a relatively easy object to craft, but a hurdle maker has to be on top of every part of their game to produce something that's even remotely fit for purpose. Selection is everything, and understanding your material to such an extent that you know the critical point between weight and strength. Use rods that are too thick and the hurdle will weigh too much to hold itself together and will take its toll on the shepherd who has to lug them around. Too thin and the hurdle will fail to contain the stubborn sheep that spies fresher, greener turnips on the other side of the pen. The zales are usually made from one inch to an inch and a half rods, four feet in length and split down the middle so that in profile they are semi-circular. Necessarily, they are tapered at the base to drive them into the pilot holes on the mould. The end zales are sometimes called shores, and these are usually left in-the-round (unsplit) and driven into the pilot holes to a deeper depth than the zales. Once the hurdle is removed from the ground, it's these shores that will be driven, like stakes, into the ground to prevent any lateral movement under pressure (from wind or sheep) at the base of the hurdle.

The first rods to be woven through the uprights, usually no thicker than an index finger, are also left in-the-round and therefore need to be slightly thinner in diameter than those that will be split to make the main body of the hurdle. These first wefts are often called spur rods, as the manner in which the hurdle is started requires a number to be laid down diagonally across the mould – spurred out – between two zales in preparation for weaving the base. This is where the specific knowledge of the hurdle maker comes into play, the canny trick of getting these rods to lock into each other through a process of interweaving that requires a very specific order. You don't just weave them through the zales and wrap them round the shores, you also interweave them with each other in an over-and-under technique. This is essential so that the lowest rods don't drop off the hurdle as it's being moved around. It's a simple bit of know-how, but without it you can't make a hurdle. These rods left in-the-round will be used for the first seven to ten inches of the hurdle's height; they have the further function, as a result of being bound around the shores at either end, of holding the whole thing together at the base. It's the technique required for this that represents another point at which a very specific skill is needed: a knowledge of the hands that can only be achieved through experience and familiarity. As I've learned to my cost, when the rod you're weaving gets to the end of the hurdle, if you simply folded it back on itself to return to the weave it would snap, not cleanly, but in a frayed splinter-like manner. The trick here is to twist and fold at the same time. This subtle sleight of hand opens up the fibres of the wood to create an almost rope-like quality that will take the hundred-and-eighty-degree change of direction that the rod must make to return to the weave.

Some hurdle makers will then use split rods to make up the main body of the hurdle. These, because essentially flat on one plane, will all be woven with the barkwood facing away from the hurdle maker, and this bias is in part responsible for the shallow arc of the pilot holes on the mould. Once stacked horizontally and compressed under the weight of logwood timber, as the thicker part of the split rods seasoned the curved tension would cause them to lock together. But old-school hurdle makers will include what's called a 'twilly hole' at a point roughly three-fifths of the way up the hurdle. This is achieved through taking in-the-round rods and weaving them in such a way that a hole is left either side of the central zale, that is, the rods are turned back on themselves at the zales either side of the central zale. These twilly rods also bind the hurdle at each end – unlike the split rods, which are trimmed short – and provide extra strength. The twilly hole is vital to the shepherd as a handle, both for when the hurdle is being positioned in the field and when they're being carried over his back – using the prongs of a thumb stick like a crutch on the exposed section of zale in the twilly hole – from stack, to cart, to field. The rest of the hurdle will be completed with split rods, but at the top a few inches of woven in-the-round rods, twisted round the end posts, are used to hold it all together. Again, a canny trick of interweaving the top rods locks them

in and prevents them from being pulled free or loosened as the
hurdle is moved around.

To watch a hurdle made at speed is a pleasure. The twisting
technique, the controlled manipulation of the rods as
they're woven, the necessary straightening of the zales as work
progresses, the tapping down into place of each rod with the
cudgel, and the masterful trimming of the split rods at each end
are all done with such speed, fluidity and precision it's almost
like watching a concert pianist at one with her instrument.
The position of cudgel and trimming bill slowly migrate with
the work back and forth along the length of the hurdle, never
out of reach, always at the same stations relative to the hurdle
maker's body. It's as if the hurdle maker could operate blind. Is
this, then, the *cræft* of hurdle making? Or is it more than that?
For me, everything about the process smacks of intelligence, but
this knowledge, ability, skill and virtue arises out of a number of
aspects and not just the actual making.

In the first instance, the landscape must be managed in such
a way as to produce good quality materials year on year. You
do not manufacture in isolation from your resource base. I find
it staggering that, despite the efforts of a committed few, such
amazing sources of renewable energy are being allowed to fall
into rot and ruin for want of basic maintenance. What I liked
most about the hurdle-maker's camp was the ephemerality of
the archaeological footprint that the process created. This was
a low-impact workstation par excellence, a pop-up production
line, hot-desking in the most convenient place from which to
source materials. A hurdle maker went into the woods with a
bag of tools, and the finished product came out.

It was also the arrangement of the work area that fascinated me. A natural ergonomic design saw material flow through the site from raw, coppiced to finished form. While the hurdle mould required the maker to kneel, the gallows and chopping block meant that he could stand up at regular intervals, adjust his position, stretch his back and shoulders and use an altogether different set of muscles. A far cry from the repetitive-strain-inducing workstations that characterise the production line of mass manufacture. And finally, of course, there is the logic of the object itself and how it works within a wider socio-economic system. Now that good archaeological evidence is emerging for wattle hurdles in the prehistoric period too, we can start to think critically about the adaptability and versatility of early prehistoric farming systems.

The Neolithic Revolution is sometimes referred to as the Farming Revolution, a time when societies throughout Europe experienced phases of agricultural transformation. Vast swathes of woodland were cleared and field systems set out in a process that continued into the Bronze Age. But it wouldn't have taken long before the virgin fertility of this newly broken ground would have been in need of replenishment, and it's inconceivable that early Bronze Age farmers were not aware of the restorative qualities of manure. Whether they used hurdle panels of the like recently recovered at Must Farm remains purely hypothetical. Yet by the close of the Anglo-Saxon period, at the end of the first millennium AD, it seems likely that hurdles were being used for these purposes.

The name alone is the main source of evidence, as we have no illustrations, historical descriptions or archaeological verification. We know that *hirde* in Old English was 'herdsman', 'shepherd' or 'keeper', and that the word *hyrdel* was translated by the Anglo-Saxons into Latin as *cratis* which, by reference to other uses, means 'hurdle' or at least 'wickerwork construction'.

In the tenth century, Ælfric the Homilist recounted the legend of the execution of St Lawrence in the fourth century, wherein the saint was roasted alive on an *isenan hyrdle* ('iron hurdle'). It seems clear in this usage that it was a moveable panel – a self-contained entity. At least some philological good came out of the poor chap's martyrdom. It's the mention of a *loc-hyrdel*, however, that is most instructive. The Old English word *loc* can mean lock, bolt or bar, but it can also be used to refer to a fold or enclosure.

CLOTH AND HURDLE, both products of the interweaving of their respective materials, have been central to the development of societies and economies from the prehistoric to the early nineteenth century. The importance of cloth seems so self-evident that it barely needs comment. Yet it's intriguing that the dwellers of the Must Farm site over four thousand years ago should have shared with the Nike purchasers in the early 2000s an inclination for a certain style of making – a weft and warp. Flick through the indexes of any standard historical texts on the eighteenth- and nineteenth-century Agricultural Revolution and you'll struggle to find any references to the role played by wattle hurdles. Yet in this one craft object is embedded the intangible history, or rather archaeology, of how the Agricultural Revolution in some parts of the country was able to support the Industrial Revolution in others. It was a lynchpin in a system of agriculture introduced to meet the growing food demands of booming populations in the cities of a fast industrialising nation. A simple matter of taking a craft over four thousand years old and applying it to new demands. While industrialists will argue that coal, iron and engineering created modern Britain,

I wouldn't disagree with them. But I would like to draw their attention to the weft and warp of the humble hurdle and the critical role it – and the hurdle maker through his *cræft* – played in feeding the workforce that operated the forges, mined the coal and manned the factories that turned Britain into the workshop of the world.

8

UNDER THATCH

TWO VERY IMPORTANT and life-changing things happened to me in the autumn of 2003. The first was that I met the woman who was to become my wife. The second was that I thatched a building. The importance of the first scarcely needs spelling out. But of the second, it's only in hindsight that I realise this event marked an inaugural moment in my relationship with both archaeology and craft. This was no ordinary thatching project, but what you might call an experimental thatch using traditional materials in a traditional style, and it opened my eyes to so many aspects of past making, not just manual dexterity and ingenuity but levels of sustainability, resourcefulness and resilience – *cræft*. I became a little bit obsessed with thatching, and for a period of about two years in my late twenties I harboured an ambition to become a thatcher – not just a regular jobbing thatcher, but a historical thatcher. It seemed to me the perfect job. I could spend my days outside, with the breeze blowing through my hair, whistling away as I engaged with one of the most ancient crafts known to man.

I should acknowledge that my interest had been greatly encouraged by a chap called Keith Payne, a thatcher from Somerset, and one of the youngest ever thatchers to qualify as a master. Then in his mid-forties, his skill and expertise were matched only by his passion for the history and archaeology of thatch and thatched buildings. It was through our shared interest that a window was opened onto the broad range of methods and materials that fall under the umbrella term 'thatch'. In popular perception it conjures up images of picturesque English villages and ancient country cottages draped with a thick mop of wheat straw tidily combed out to make a smooth and neat finish. But in the origins of the term itself is a much wider definition, one that broadens the story of thatch from the cosy English rural scene into a more global and all-encompassing story of mankind.

Our modern word is derived from the Old English term *þæc*, which itself probably comes from a more ancient etymological root shared by the Latin *tectum*. In both these instances it's used to describe a roof; a covered structure, roofed enclosure, shelter, house, dwelling or abode. So, in many respects, when it comes to thatching, in ancient times at least, anything goes. I think there is probably a distinction to be made from the later twelfth century between thatch and the use of fired ceramic tiles, slates and split stones, because at about this time the density of occupation in cities was reaching a point where it was becoming dangerous to combine large volumes of suspended desiccated organic material with industries reliant on open furnaces. Major conflagrations are recorded for London in 1077 and 1087; in Canterbury, Exeter, Winchester and London during the particularly hot summer of 1161; Glastonbury in 1184, Chichester in 1187 and Worcester in 1202.

The situation had become so bad in London that in 1212 an ordinance was passed that prohibited any future covering of roofs in thatch, and stipulated that existing thatched roofs had to be given a fire-retardant limewash. In Norwich, after a succession of major blazes throughout the fifteenth century, by 1509 roof tiles had become compulsory. But outside of the cities it was only really buildings of higher status, such as churches or manor houses, that would have been roofed in more durable materials. Otherwise, thatch and thatching were ubiquitous, and this persisted up until the middle decades of the nineteenth century when the installation of steam railways facilitated the national distribution of heavy loads of slates from Wales and fired tiles from the clay lowlands of England. Even today, as you look at a distribution map of existing thatched cottages, you can see how pockets of relatively numerous thatched buildings lie in places remote from the national rail network.

So, in these early days, before the industrial quarrying of slate and firing of tiles, what you used as your thatching material, and how you chose to fix and fasten it to your roof, was very much open to your own interpretation. There were no hard and fast rules, and this is what makes thatch so *cræfty*. There is no greater demonstration of *Homo faber*'s intelligence and resourcefulness than his ability to build a watertight and robust structure over his family's heads using entirely organic materials sourced from the immediate environment. The early training I'd been given by Keith triggered in me an exploration of the thatching tradition in Britain and I was amazed at the rich variety of materials and methods used to cover dwellings and outbuildings. From wheat straw in the south of England, water reed in the east, marram and sedge grasses and even seaweed in the Highlands and Islands, to heather and gorse in the uplands of northern England and southern Scotland, all manner of plants were being fastened through a range of ingenious techniques to create waterproof coverings of relatively lasting stature. Furthermore, as an archaeologist, I realised that many aspects of the social and economic history of a people, their buildings and their landscapes could be deduced from the style of thatching and the choice of materials.

Take, for example, the case of a smallholding built in 1880 at the crofting township of Locheport on the island of North Uist in the Outer Hebrides. Here, in a complete overhaul of the roof, a section was cut through the thatch exposing layers representing no fewer than five episodes of thatching, each signifying a period of around twenty-five years. For a part of the world that has an average wind speed of over 97 knots per second and an average rainfall of over 1,000 millimetres per year, the durability of these thatches is testimony to the skill of the Hebridean thatchers. But what was really interesting were the changes over time in the materials and the techniques. The first phase (when the smallholding was built) comprised a turf basal layer overlain with a basecoat of mixed cereal, marram grasses and rush, and then finished with a topcoat of clean cereal straw. The whole lot was fastened down with twisted heather ropes that would have been weighted with stones – a remarkable use of native materials. For the second phase of thatching (c.1905), there was a change in plant material represented by a thick layer of heather, also held down with heather ropes. A third phase (c.1930) consisted of a similarly thick layer of heather but in this instance there was no evidence for it being secured by heather rope. It's most likely that at this stage in the building's history either a second-hand herring net or a fine-gauge chicken wire was introduced as an innovative way to secure the material to the roof and to deter rodents and birds. The two final phases (c.1955 and 1980 respectively), consisted each of a basecoat of bracken and a thinner topcoat of heather.

The sequence of thatching types told an interesting story and one that tied in closely to the recent history of the island and the relationship of the Outer Hebrides to the wider world. While heather is undoubtedly a more durable thatching material, there is no point in undertaking the arduous and time-consuming job of cutting a roof's worth of heather if you already have an abundance of cereal straw lying around the farmstead as a by-product of growing barley, oats and rye. It may have been that the cereal wasn't up to the job of providing protection against the harsh Atlantic storms, but the change to heather at around the turn of the century might also be a reflection of the changes in farming practices on the islands. Perhaps a little later than elsewhere in Britain, arable farmers on Uist would have been feeling the pinch of the emerging global trade in cereal crops facilitated by improvements in steam-shipping and the opening up of the American prairies to the growing of vast acreages of wheat. Such was the deluge of grain finding its way across the Atlantic that it became economically unviable to grow cereals on small island plots when you could import in bulk at such low prices. So, without access to a ready supply of cereal straw, and in a time of financial uncertainty, it may be that the crofters returned to the craft of their ancestors and to the abundant heather that grew around them on the island.

By the middle of the twentieth century, further developments in agricultural practices, spurred on by the impacts of two world wars, would have more profoundly influenced craft practices, even in this remote part of Britain. The introduction of intensive sheep farming would have reached its peak at about this time, and the grazing of large flocks would have impacted on the availability of good heather. Thatching requires tall single-stemmed plants that grow thick among themselves and competing for the light. The effect of grazing animals is to prune the plants to such an extent that they bush up tightly against the ground. It may be that the local heather supplies had never recovered from the harvesting for the second and third phases of thatch. From the sample taken, it would appear that the plants were pulled rather than cut, significantly limiting a heather colony's capacity for regrowth and reseeding. Equally, bracken tends to be one of the first invasive arable weeds to creep into abandoned plough soils, and it might have been the case that by the late 1940s fields that had once been kept in a state of neutral acidity through routine kelping (the manuring of alkaline-rich seaweeds) and used to grow crops had been largely commandeered by this acidic-soil-loving plant.

In this small section of thatch was an interesting and important story, and one that was just as well placed as any written account to tell the tale of mankind's interaction with this harsh but majestic landscape. Each generation of thatcher, perhaps with some authorship over their choice of materials, was as much constrained by the greater current of economic events on the world stage as they were by island tradition. The sequences of differing thatches provide an alternative perspective on essentially the same story. This wasn't a verbal or textual response to wider historic events but a physical, practical and craft-based response. It was a craftiness on the behalf of the island crofter-thatcher that enabled and facilitated human existence on these remote islands, buffering against the force of the North Atlantic winds and equally against the rising storm of a global economy.

The sequence at Locheport provides an excellent example of how a craft should never be considered in isolation from its immediate surroundings, and that the resourcing of a craft is almost as important as the end product's functioning value. What works in one part of the world might not be the chosen method in another because wider social and economic factors take precedence over technical superiority. I saw this most vividly during the course of undertaking my first thatch. This is when I met Keith, who was to be my thatching mentor as I tried to recreate an authentic sixteenth-century thatched roof for a lowly cattle shed. Keith was a very laid-back individual, always smiling and joking and quick to burst enthusiastically into explaining some tiny detail concerning a particular aspect of thatching. He was the classic example of the relaxed attitude and self-assurance that come from spending one's life perfecting a craft. On my day of arrival on site he informed me that we were going to use what was called a stubble thatch of wheat straw, overlying a basecoat of bracken, and that we would use hazelwood spars, gads and liggers to secure the thatch. Like most crafts, thatching has a whole host of terms assigned to the tools and techniques of the trade, and as work proceeded I very quickly had to get my head around biddles, yelms, stobs, stolches, whimbels, spuds and leggats. It was a steep lexiconic, as well as physical, learning curve.

Interestingly enough, in France the practice of stubble thatching is preserved in the French *chaumière*, a word for a small shelter, derived from *chaume*, meaning 'stubble left in the field'. In England there is no such etymological survival, and it may be that the widespread adoption of corrugated tin roofing in the more industrialised farmscape of Britain sounded an earlier death knell for what was likely a commonly used means of roofing outbuildings and temporary shelters. It could also be because Britain, during the golden age of farming in the 1850s, adopted mechanical harvesting equipment on a scale not witnessed in France until the interwar years. Thus, from an earlier period, the material required for a stubble thatch was no longer as widely available.

Today, we turn the ignition key in a combined harvester and run it up and down the field cutting the crop, threshing it and separating the wheat from the chaff in a single – combined – mechanical operation. But before the twentieth century, these processes were undertaken independently of each other using a range of different methods. Modern varieties of cereals have been bred differently from the older varieties and have, as a consequence, a much shorter stem than the ancient varieties, which would routinely grow to heights of five feet and over. Even by the nineteenth century, many harvesting methods had already become mechanised so that horse-drawn reapers would cut these tall-stemmed cereal plants, while an army of labourers bound them up into sheaves before 'stooking' them – stacking them in small groups – to dry in the field. Collected in, these sheaves would at some point later in the year be passed through a steam or horse-powered threshing drum and the grain separated out from the plants. During this process, the straw of the plant would have taken a bit of a beating and, if its intended use was for thatching, it was combed back into straight sheaves, either manually on a combing frame or through a mechanical combing box. All very labour-intensive, but not as much as doing all these processes by hand, which, at the beginning of the nineteenth century, was the case for farms and farmers the world over.

It's only when you're doing things by hand, and don't have the speed and power of machines to rely on, that you start having to get crafty with the way you set about the work of harvesting cereal crops. It goes without saying that harvesting without machines is a much more time-consuming process, and as a consequence you have to factor in a range of variables to which the mechanised harvester is largely oblivious. You are, for example, much more at the mercy of the weather. In a full harvesting season you may have to alter the methods for bringing the crop in; cut early in some places and leave it late in others. This has an effect on how you harvest a crop and how you process it. A crop cut early may still need to do some drying in the field to finish off. In which case, you might want to cut it with its stem on and stook the sheaves, leaving them to dry in what's left of the summer sun. For this job a scythe would be used. Such implements are known from the early Middle Ages across Europe. With a long shaft, the operator can stand upright and swing the blade through the base of the plant, severing it as close to the ground as possible, then the loose plants can be bunched up by hand by an army of bunchers and binders. This process results in a short stubble left in the field.

But you might leave some of the crop till late in the season – perhaps because of a shortage of labour or because a bout of summer rain has suspended the harvest. In this situation, with the plants reaching maturity, it would be dangerous to swing a scythe at them. With the grains much looser in the ear, jarring the plants at their base with the blade would run the risk of dashing the grains from the heads and losing much of the crop to the ground. It might be better to use a serrated sickle and, taking hold of a handful of plants below the ears, draw the sickle around the stems in a more controlled fashion, and cut the grain-holding ears in bunches. It's a more delicate and time-consuming operation but, for our purposes, it tends to leave a much longer stubble in the field, which must then be scythed down and bound separately before the field can be ploughed for the next crop. In some ways, this stubble is a substandard thatching material by virtue of the fact that by the time it finally makes it into store it's shorter in the stem. But this is what made it interesting to Keith: while he'd spent his life thatching people's houses, for which only the finest materials would be used, he'd never thatched a low-status building working with substandard materials. It therefore represented a challenge to explore the efficacy of certain coat thicknesses as well as methods of securing and binding. For my own research purposes, this led to estimating the intervals at which buildings would need re-thatching, and thus the ubiquity of thatching skills as a currency among a rural community.

Perhaps the biggest insight our stubble thatch project gave me was in the sourcing of materials. The bracken basecoat had at first seemed like one of the simplest parts of the process as the hill farm where we were based was almost entirely surrounded by the stuff, growing anywhere it was allowed to on the precipitous slopes of the valley. My first thought was how fortuitous it was that we were picking in September. This was indeed the perfect time of year to be sourcing the plant as it was at its largest and still green in the frond. Any earlier in the year and we would have had to pull many more plants to get the right bulk; any later and we would have been working with a drier and more brittle substance. Bracken basecoats were, I concluded, an early autumn entity and would have been negotiated around any late cereal harvests and before the main fruit harvests.

We soon found that cutting with grass hooks and gathering up in the hand was more time-consuming than pulling the plants by hand and laying them in an orderly manner on a sledge. Certainly, this method did more damage to the plant than cutting as it severed the root connection and wounded the rhizomes. Bracken is sustained by a rhizome, a carpet of root mass that creeps its way through the subsoil, spreading its tentacles into the broken ground. Our English word is said to derive from the Old Norse *bracken* and the Swedish *bräken*, meaning 'fern', but I also entertain a derivation from the Old English *brecan*, meaning to 'break' or 'burst', a reflection of how the species multiplies itself and spreads forth most easily in *broken* ground, land disturbed by either the forces of nature or interference from the plough. We were, however, winning some fans among the local farming fraternity, one of whom, on an afternoon walk past one of our favoured picking places, commented on how Wales was losing 1 per cent of its rural areas to the encroachment of bracken every year.

After about three days it was beginning to feel distinctly like we were going some way towards mitigating this 1 per cent – and still had another half of the roof to go. We became adept at spotting particularly thick stands of the stuff, and even on our way to and from the site in the morning and evening we'd pick a sledge's worth. Only when our backs were broken and our hands raw (even through thick leather gauntlets) did Keith feel that we had enough to begin the thatch. This, then, gave me my very first insight into the matter of resourcing – factoring in the effort and energy required to bring together enough material for the job. I'd become so used to ordering up materials in bulk and having them delivered to site that I had no real knowledge of the labour involved in physically procuring enough material myself.

I'd experienced a similar situation with the sourcing of the rafters for the roof superstructure. Here, though, it was less about bulk and more about quality. I'd spent about a week in the woods cutting straight-growing hazel and ash poles from coppiced stools and young stands. They weren't all perfect but were the best I could source from the limited woodland available. I realised that the ability to match cunning resourcefulness to nature's variables would have played a key role in the *cræftiness* of the seventeenth-century barn-roofer. So I had to work with what I had and match up each rafter with a similar-looking pair and alternate between good pairs and not so good pairs, spreading the quality evenly across the roof. Within each rafter peg holes were manually augered and through them home-made oak pegs, all finished on a shave horse, were used to secure the rafters to the wall plates (strong timbers along the stone walls of the barn) and the purlins (a horizontally set timber that crosses at a mid-point up the roof to support the rafters on their trajectory towards the ridge). Square pegs were made for round holes on the theory that they naturally lock together as the green wood of the rafters seasons and tightens up around the already seasoned hardwood peg.

We then used an ingenious method – Keith's idea – to create a grid on which to thatch. Contemporary approaches would involve taking machine-cut batons and nailing them to the rafters. This wasn't an option for us. The rafters I'd sourced gave the roof, shall we say, an organic feel. Even if one man's flawed irregularity was another man's vernacular charm, the discrepancy between one rafter and the next was such that it would have been impossible to use straight-cut batons. Instead, coppiced hazel rods were sourced – another week spent in the woods – and used to create an open weave between the rafters. The resulting superstructure was therefore half pegged and jointed traditional carpentry framing and half basket – and the benefit was that the woven material reinforced the structural integrity of the rafters. No nails, no screws, and not a machine-sawn baton in sight.

The bracken went on in 'stolches' rather than courses, meaning columns rather than rows. It needed to go on much thicker than I'd envisaged and was secured with twine cord and 'gads'. These were short lengths of hazel rod about two feet long, which had been split down the middle. The job required two men: one inside the building behind the rafters and the other on top of the roof. Outside, the thatcher would tie one end of the cord to the gad and would then, using a thatcher's needle, thread the other end of the cord through the thick mat of bracken. It was the job of the thatcher on the inside to receive the needle and pass it back through the roof, having passed it round two or maybe three rafters. Receiving the needle on the outside, the cord would then be fastened to the gad using a slipknot that could be tightened as the gad was pressed down with the full force of the thatcher's knees to compress the bracken into a compacted mat. The better the compaction, the better the foundation for the spars that were going to be used to secure the topcoat.

Once our woven roof had been covered in a thick bracken mop, the ridgeline, gable ends and eaves were built up with straw 'bottles' – ready-bound handfuls of straw thatch. These were the exposed edges of the roof and therefore needed extra enforcement. Then came the stubble thatch. Unfortunately for us, this was raw material that wasn't readily available from present-day thatching suppliers – and probably hadn't been since the early nineteenth century. So, to simulate the length of stubble-thatching material, we used hedging shears to cut the ears from a lorryload of sheaves that we'd had delivered. This was hard, monotonous work, but a darn sight easier than doing it in the field as it had been centuries ago.

The topcoat went on, like the bracken, in stolches. Keith did one side of the building, demonstrating and talking me through the process in his dulcet West Country tones. Every action, every sleight of hand builds towards the crafted product. There is no one particular moment or movement when the craft skill is realised; it's a slow and incremental art. Every aspect of securing the stubble had to be done to a standard. In the first place, the thickness of straw had to be even. Any irregularities in bulk would quickly be exploited by rainwater run-off, which would naturally find itself flowing to thinner areas and creating gulleys, which in turn would further eat into the roof. This coat was secured with a handful of straw – a 'bond' – that was set horizontally and secured at each end with a spar. Spars were like staples, made from lengths of hazel rods roughly the same size as the gads, only this time the rod was split into quarters with each end then being sharpened (like a pencil) before being twisted back on itself. Fortunately, I didn't have to source these. If I had, I'd still be there now, wandering the hillsides of south-east Wales, desperately trying to find enough straight coppiced rods to make up the three thousand spars that were required for this modest farm building. At this juncture, I was truly beginning to comprehend the sheer level of work that went into a simple barn roof. Not just its making but its resourcing.

Each spar needed to be driven in at the correct angle, hard enough that the fingers couldn't draw it out again. Each new layer in the stolches needed to overlap just the right amount to cover the binding of the previous layer but also maintain enough bulk of material to keep the pitch of the thatch aligned with the pitch of the rafters, again ensuring that a uniformity of thickness was safeguarded at all times. This craft was all about the meticulous adherence to a standard, not just with every single fastening and laying of straw but with every single movement that was made on the roof. Every turn, every carry, every placement of tool, every manual shaping of material had to take place within a flow of kinaesthetic sensibility. This was clearly a major part of the *cræft* of thatching as it had been over a century ago, but it simply cannot be considered in isolation from the resourcing of the raw materials and the intelligence that comes with taking local organic matter and converting it into a functioning entity.

Everything on that roof could have been sourced from the farmstead. Admittedly, we imported some commercially sourced thatching straw, modified for the type of thatch that was required, and we used a flax twine to secure the gads and bind the bottles. But we could just as well have used lengths of bramble cane or honeysuckle vine shoots, which may have had a longer lifespan but would have taken an extra week to source and prepare. Even taking this minor shortcut into account, on the evening that the barn was finished, taking in the whole completed spectacle for the first time, an enormous pride welled up inside me at what I had helped to achieve. In so many ways this roof had been a gateway into a new world for me. My life would never be the same. A new Alex emerged. Archaeology became so much more than just stuff in the ground. It became an exploration of what it was to be human, not only because we are makers but because we are resourcers, gatherers with an inveterate knowledge of the natural world around us.

I HAD WELL AND truly caught the thatching bug. My wife and I couldn't even go on an afternoon's ramble without me picking various plants and considering their potential as a thatch covering. I visited the Highlands on a thatching tour of old Black Houses; I experimented with different thatch types, covering outbuildings with nettles, privies with oat straw, sheds with dock leaf plants. I was impressed by the properties of tansy, a tall perennial weed with a yellow flower and thick woody stem; pulled rather than cut, it performed admirably as a covering over a home-made chicken run. I was unimpressed with Virginia creeper; voluminous and flexible to begin with, its pithy heart rotted out too quickly, leaving its wafer-like barkwood to crumble under the touch. I was most interested in the creative styles of roofing that were adopted in the more extreme parts of Britain and Europe, especially in places where it was hard to come by good serviceable timber to build the roof in the first place.

The seaweed thatches of Læsø in Denmark were a particularly interesting way of making use of a material that was abundant on the beaches of the island. With the driftwood timber sourced from the same locality, and both having been heavily impregnated with saltwater, these roofs preserved well and thus had a remarkably long lifespan. On the Hebridean islands of Scotland the shortage of good straight timber was also keenly felt, and any substantial driftwood that washed up on the beaches would immediately be dedicated to the role of roofing timber in buildings that were necessarily squat to resist the winter gales of the Atlantic Ocean. But it was the method of replicating rafters and batons to hold the basal layers of thatch that was most creative. In place of cut timber, the islanders used a twisted heather rope that was secured to a timber flush with the wall plate. It was then passed up over the ridgeline and down the other side of the pitch before being locked around another timber on the other side and passed back up towards the ridge. This created a giant net across the roof onto which cut turves were laid, overlapping like tiles and with a marram grass thatch secured over the top by virtue of another heather rope net. This really was ingenious and something that I just had to see for myself.

Fortunately, I didn't have to travel as far as the Hebrides. Keith, aware of my obsession for intriguing thatches, called me one day and said he had a particularly interesting job up on the coast of West Wales and that I should drive over and take a look. This wasn't an offer I could turn down easily and, persuading my wife into taking an impromptu break, I duly packed the car for an extended stay. The building Keith and his team were working on was an eighteenth-century farmhouse that had retained a number of its early features, chief among them a thatched roof. While the main part of the farmhouse had been thatched in a conventional style, the roof of the byre was altogether more imaginative in its construction. In a curious paradox, the thatch had survived by dint of the technological development that had killed off the vernacular thatching of outbuildings across the country: corrugated tin.

As British farming became increasingly industrialised towards the end of the nineteenth century, corrugated pressed tin-plate roofing panels became much more widely available. At the time, the take-up of this innovative form of making shelters was slow. Even today, farmers remain deeply conservative by nature, but when you had a steady workforce of farm labourers with ingrained thatching skills to hand, there was even less incentive for newfangled investments. But all this was to change with the human cost of the First World War. Robbed of a generation of young farmhands, and a resident rural workforce increasingly replaced by mechanised traction, farmers and landowners needed to find a more permanent means to keep cereal crops, hay and livestock dry. Corrugated tin was to provide that lasting solution, and with the need for increasing cover to protect vast increases in home-grown produce during the Second World War, as well as a the huge demand for temporary housing, barracks and prisoner-of-war camps, its rise to prominence was complete. So the age-old skills of thatching in an ad hoc fashion – to quickly cover, with local organic materials, an outbuilding, a hayrick or an old tithe barn – died out virtually overnight. It wasn't just thatch that suffered. As older peg tile and slate roofs became tired and in need of repair, the cheapest option was to source a lightweight alternative, changing both the visual character and the skill set of a farmscape in a single swift move.

For the farmhouse in West Wales, this meant that the byre, instead of receiving a fresh coat of thatch in the traditional local style, was boarded over with tin. But remarkably, almost as if in recognition of this moment of profound change, the workers applying the tin had secured it directly on top of the existing thatch. As a consequence, it had become desiccated and left in a state of suspended decay for the best part of a century. So when it was exposed, as part of a redevelopment of the farmstead, it was immediately apparent to the developers that this was a unique and incredibly rare survival, and that steps had to be taken to ensure that it was, at the very least, replicated. In stepped Alex, the trained archaeologist, to examine in scientific detail the sequence of historical thatch. And, for the fanatic I had become, this was a juicy one.

It was heavily affected by rodent action, and as soon as all the tin was removed we could see rat burrows criss-crossing the surface and diving behind the topcoat. In the excavation of just a stolch's worth, I pulled out four mummified rat carcasses. But once the disturbed material and rodent detritus had been cleaned out, the shape and form of the roof could be clearly examined. And it was a beauty. I marvelled at the substratum. Here, straw ropes were used in much the same way as the heather ropes in the Hebridean tradition to create a net on which to place the basal layers. So, in total, the structural timber of this roof amounted to a mere seven poles, all set horizontally in the gable end of the farmhouse proper and all supported at their other end in the external gable wall of the byre itself. One served as the ridge pole, two as the wall plates, two as purlins and the final two as locking purlins – timbers the straw rope was locked around to gain tension before being returned over the ridgeline.

On top of the straw ropes the most ingenious of basecoats had been laid. Here, relatively mature gorse bushes (say, four to six years old) had been placed face down onto the straw ropes. Gorse is a very prickly customer and being laid in this way it did two things: first, the prickles acted as a deterrent to rodents, protecting the structural straw from any unwanted gnawing; second, they helped the gorse bushes lock into the ropes, creating enough friction such that no extra fastenings were needed. Solidity in this layer came from the root boles of the bushes, which had been cut out of the ground in a way that a substantial and very thick turf mat had come out with each plant. They were like turf building blocks but with an added rodent deterrent. Very clever indeed. But things got better, because into this thick earthen layer the topcoat had been 'stob' thatched. Stob thatching involves taking a handful of thatching straw, twisting its end back over onto itself and stobbing it – thrusting it with a stob stick – into the earth as a means of primary securing.

As a source of secondary fastening, rows of assorted pegs appeared to have been hammered into the external surface. It was difficult to know what form this peg fastening had taken across the whole roof as the butt ends were so badly nibbled by rats, but where they did survive they were of all shapes and sizes: irregular, curved, knotted and of varying thickness. These pegs were very definitely not like the uniform commercially sourced spars that we'd worked with on our seventeenth-century barn reconstruction. These were makeshift. They were 'hedgerow'. The use of these pegs was a classic example of playing with the hand that nature deals you. This whole roof was a window into true *cræft*.

It was amazing to think how all these materials, so alien to us today, could be made to work together to produce a functioning roof. This was so very far from a trip to the builders' merchants today. This was an entirely different world, an entirely different way of thinking about structure, about building and about materials. And it can't have been that long ago – maybe mid-nineteenth century at the latest – that this roof had been put up.

But how old this style of thatching was, how old the knowledge, the *cræft*, who knows? As an organic material, thatches almost never survive in the archaeological record, and when they do it's near as damn impossible to say for certain that an excavated organic-rich deposit was once a roof. So, we just don't know how people thatched in the past.

To begin with, before my obsession took hold, I'd never imagined that you could build the majority of the supporting superstructure of a roof without timber. I think I calculated that to have used rafters to support the basecoat of this humble Welsh byre would have needed around thirty to forty poles on top of the seven already employed. What do you do when you don't have access to this kind of material? You take an everyday material, a by-product of your annual harvest, and you twist it like a yarn, incrementally feeding in more and more straw as you proceed. When you're happy with the length, you double it over on itself and twist again, thereby locking one yarn against the other to make a twine. And thus, you have a rope. Simple. What this method of thatching proved to me most forcefully is that it's not that we have lost these ancient skills, it's worse than that. It's that we have lost the conception of these skills and what they can do for us.

A FEW YEARS LATER I got my own chance to build a roof without having access to good structural timber. The opportunity rather crept up on me one winter's day. I found myself in a situation where the hedgerows on the farm were completely out of control and in dire need of care and attention. While I was familiar with the key principles of hedge maintenance, I was somewhat overawed by the scale of my hedgerows. A colleague looked up some local hedging talent – one a recently crowned national champion – and with very little persuasion they agreed to head over and spend the day offering me tips and advice. Hedging is probably my most enjoyable winter pastime. There's no truer way of getting hands-on with the landscape than sculpting the form of a living stock-proof barrier, manipulating the inherent strength of thorn and hazel and stepping back regularly to take in the line and uniformity of one's work. But on this occasion I was treated to a masterclass.

The most important lesson was in selection: which branches or shoots to keep in the hedge for laying and which to dispense with and cut out. Previously, I'd always been reluctant to cut material out. After all, I reasoned, if you wanted a cattle-proof barrier, surely the more branches it had the more impenetrable it would be? So the biggest eye-opener of our communal day of hedging was the huge amount of material removed. The next morning, as I lay in bed and wondered what on earth to do with this large pile of unwanted branches, it occurred to me that the reason those hedgers had been so jovial all day was because they knew they wouldn't have to clear up. I rolled out of bed, flexed my stiff shoulders and rubbed my aching elbow. It's a measure of our bourgeois times that such pains are known as tennis and not hedger's or carpenter's elbow.

This was a job that had to be done now. Today. Despite my aches and pains and a long week's worth of work behind me. Why? Well, in part because the puritan in me felt that it didn't seem right to have had so much fun yesterday without there being a catch. But also because the longer I left it the more likely it was that the grass in the field would grow around the discarded branches making it difficult to see what needed clearing. If left long enough it would impede grazing. We'd also, in places, cut some substantial trunks out of the neglected hedge, which could have a disastrous effect on the delicate blades of the mower should the field be selected for haymaking. There was no getting away from it: I was going to spend my Saturday hacking, chopping and sawing unwanted hedgerow branches and brash (the finer twigs and branches) into manageable lengths and finding a way of disposing of them. As I left the cottage, I stuffed some bread and cheese into my jacket pocket, grabbed my billhook and bow saw from the toolshed, tied a hessian sack around my shoulders to keep off the worst of the rain, and headed out to the fields.

When I arrived I was somewhat dismayed. The row of discarded branches of all shapes and sizes seemed almost as large as the hedgerow itself, and another pile of the same size awaited me on the hedge's other side. Of course, the quickest way to deal with this mass of wood would have been to create a loosely built fire, chuck a cup of diesel on and send it all up in a ball of flames. This would have taken maybe a couple of hours, with perhaps an evening revisit to rake any unburned brash over the embers to finish the job off. But since waking I'd been toying with the idea of doing something else – something very old-fashioned – and it was for this reason that I'd dropped by the farmyard on my way over to pick up a ball of baler twine. The job I had in mind was undoubtedly going to take me longer, but this was my chance to build a thatched roof without using any major structural timbers. To do it, I would have to make something else in the process – the necessary building blocks for the substratum of my roof. These building blocks were a commodity I would struggle to find a use for in the twenty-first century, but a mere 150 years ago they would have been indispensable to rural life and the effective running of the farmhouse and cottage kitchen, with any surplus fetching a tidy price at the local wood fair. So, to make my thatched roof, I'd decided to convert this pile of dishevelled hedgerow brash into faggots.

I'd first learned of faggot-roof construction from an old Somerset thatcher many years ago when discussing different types of temporary roof coverings for wood stores. My question was simple: what did we do in an age before polythene tarpaulins or corrugated tin to keep the rain off our firewood? The faggot roof was an interesting idea and was logical in its construction and function: instead of cutting lengths of timber to use as rafters to build the pitch of the roof – a job that involves good timber and complex carpentry and joinery – just bank up a pile of faggots in pyramid form on top of your stacked firewood. This created the substructure of the roof which, when lashed down, was covered with a topcoat of thatch. I've since discovered late nineteenth-century photos of Devon farmyards in which you can see wood stores that have been constructed using this exact method. And it figures that, once the firewood is well seasoned, the faggot roof can be incrementally removed along with the firewood as both are required throughout the winter months.

The loose definition of faggot is a bundle of firewood. The name derives from the vulgar Latin for a 'bundle', *facus*, and the word faggot first appeared in the English language in the medieval period. Early records show that faggots intended for sale were to be made to a particular length and girth. But in reality you made your own faggots to fit your own needs, particularly the size of your fire or oven. Sometimes the nature of the brash also dictates size and shape, but it's in the effective making of a faggot that its usefulness as a piece of fuel resides. To build the wood-store roof I was going to need at least a hundred of them, which was no mean feat for a beginner. It was in the process of making these faggots – and in my subsequent years of dealing with the brash produced by various episodes of coppicing, hedging and felling – that I developed my own technique.

As with any job, it's always most productive to work in pairs, and persuading someone to help exponentially speeds up the whole process. My advice is to make a first pass through your brash material to sort the branches into appropriate thicknesses.

It doesn't hurt to have a few larger pieces of wood in there, but anything too big should be sawn to length and consigned to the firewood stack. To make the faggot you take the thick end of a branch and fold it over at the required length and keep doing so until you run out of branch. With the second branch, follow the same procedure but place the thick end at the opposite end of the faggot from the previous branch. This will enable you to keep an even thickness throughout.

Occasionally, you're confronted with a secondary branch of similar thickness – still attached to the main body of the first – and this can either be removed completely for folding in from scratch or it can be doubled back on itself and folded in from that point. The best way of folding a branch is to twist it as you bend or fold; simply folding a branch at intervals will result in a spring-like quality and it will pop out into its original form as soon as you release it. Twisting as you fold helps mitigate this, and occasionally locks the various folds in on each other. This is where a second person is of most benefit: they select a comfortable location to sit, as the designated faggot-maker, while the first person prepares and trims the branches, sorts out the wood into various thicknesses and feeds them the raw material. The benefit of being seated is that the folded lengths are secured on the lap of the faggot-maker, allowing them to get that crucial compaction while successive bunches of twist-folded branches are added to the faggot to achieve the required girth. Once you're happy that you have a faggot of the right size and under the right compaction, it can be lashed with twine at both ends. It stands to reason that the best faggots are the ones with the most amount of thin wood packed into the specified size.

The thickness and compaction of a faggot is not something you tend to think twice about until you find yourself relying on them on a daily basis. But in an age before piped gas and electric heating elements, it would have been inconceivable that the pile of wood created by my hedgerow maintenance would be converted into anything other than firewood, and faggoting was the best way of handling and transporting the smaller pieces of wood. It would have been a necessary part of any agricultural community's farming year because all the farmhouse, dairy, brew-house and cottage fires would depend on this crucial resource. But faggoting was about more than just making use of scraps of wood to create a burnable log shape. They were as indispensable as bulky firewood, being the only means fires could be easily started. I admit they're time-consuming to make, but you only realise the true value of a faggot when you return home late on a midwinter's day after an extended trip away. The cottage is cold, you need heat, and you need it quick. Nothing warms a room as quickly as a torched faggot.

Faggots also have an advantage over a dense hardwood, such as a log of ash or oak, in that they represent a more effective use of fuel and energy. What faggots allow is a firing – a single quick burst of flame for a specific task. This is why they're so important for bread ovens or copper pots. In both cases, they generate very high temperatures very quickly. For the bread oven, a high residual heat is needed to bake off risen dough; for the copper pot, the quick intense heat can swiftly boil up a basin of water for laundry, beer-making or any other hot water needs in the household. More generally, a faggot is used to fire a hearth – the communal fire in the cottage or house. Very often the members of the labouring household will have been out working for the day. Arriving home, food and relaxation would be their immediate aim before bed. Lighting a faggot in the hearth would provide enough heat to cook the evening meal and boil a kettle of water. And, as in the case of a bread oven, with the doors closed and curtains drawn, the short burst of heat from the faggot-wood fire would be enough to provide residual heat in the room for the evening.

So, I'd learned that faggots were important and that it was therefore important to get their construction right. But I had yet to see how they functioned as a roofing material. It seemed to complete the circle for me – the perfect *cræft*. Not only did faggots make use of material that in the modern age would be considered waste but they could also be used to protect and keep dry the very wood that they provide the ignition for. What wasn't to like about the whole project? Of course, it meant that I also had to source a substantial amount of thatching material. As is often the case with my experimental historical crafts addiction, one extremely long and arduous task leads to another extremely long and arduous task. But the thought of replacing the ragged plastic tarpaulin that covered my woodpile with an authentic roof covering was too much of a lure. I got there in the end. In a tail-wagging-the-dog kind of way, I had to completely rebuild the wood stack to accommodate the dimensions of the faggots and the pitch of the roof that they dictated. The bracken I used for the thatch was a bit past its best (it was late October) and I ran out about two-thirds of the way through, resorting to equally brittle nettles for the final third. But it shed water, kept the bulk of the timber dry and in the end, for one winter at least, gave me a source of instantly combustible firewood. I didn't throw that old ragged tarpaulin away, I folded it neatly into a polythene bag and stashed it in the rafters of one of the outhouses. Just in case.

9

THE SHOE AND THE HARNESS

Ic bicge hyda ond fell, ond gearkie hig mid cræfte minon, ond wyrce of him gescy mistlices cynnes, swyftleras ond sceos, leþerhosa ond butericas, bridel-þwancgas ond geræda, flaxan, þinnan ond higdifatu, spurleþera ond hælftra, pusan ond fœtelsas; ond nan eower nele oferwintran buton minon cræfte.

Ælfric's *Colloquy*, chapter 97

I buy hides and skins and I prepare them by my craft, and I make of them boots of varying kinds, ankle leathers, shoes, breeches and bottles, bridle-things, flasks and bougets, leather neck-pieces, spur leather, halters, bags and pouches, and nobody could wish to go through the winter without my craft.

ÆLFRIC'S *Colloquy* IS dated to the eleventh century and takes the form of a dialogue, between master and pupils, about the roles of various members of the rural working community and the virtues of their craft. Written in both Latin and Old English, it was originally designed as a translation aid for pupils, but for the modern reader it provides a fascinating insight into life in the Anglo-Saxon countryside. A majority of present-day craftspeople would no doubt concede that the most versatile of all materials is leather, and the list provided by the *sceo-wyrhta* (shoe-wright) in this part of the dialogue, while lengthy and apparently comprehensive by Anglo-Saxon standards, does little justice to the vast array of uses that leather has afforded us over the past thousand years. It's the last line of the shoe-wright's plea which so impresses on us the fundamental importance of leather and its significance to human endeavour since the very beginnings of complex societies: 'nobody could wish to go through the winter without my craft'.

Leather has enabled us to go beyond the limitations of our human physiology and the constraints of environmental conditions in a way that no other material has. It is to leather that we owe mankind's conquest of the planet, from the arid and inhospitable wastes of the equatorial deserts to the cold and wet recesses of the northern and southern hemispheres. Leather gave us a level of protection that, even to this day, is matched by few other materials. However, it's not simply the case that a skin can be flayed from the corpse of a freshly killed animal and immediately pressed into service as a protective garment or a functioning object. In every instance where we have evidence – be it historical or archaeological – of the use of animal skins, we must consider that a crafted process was undertaken to convert what was effectively raw flesh into something more appealing and long-lasting for human use.

It's likely that early forms of preservation were quite simple. In Book 17 of Homer's *Iliad* we're told that hides were stretched and their surface worked down and drenched in fat and gore. Parallels can be found in the practices of Native Americans, who are said to have bathed the skins in a solution of lye before stretching them out to dry and then working the fat and brains of buffalo or elk into them. The final process consisted of smoking them slowly over a smouldering fire in a sealed tent in order to cure them. Rinsed and cleaned, the oils smothered into them would help retain flexibility and a level of durability, but as a primitive measure this process didn't produce leather as we know it. In more modern forms fat and brains were replaced or supplemented with butter and egg yolk in a method known as 'stocking', where a blunt spatula was used to force these greasy materials into the hide on a set of stocks. Numerous episodes of stocking and drying would improve the final product, although the repeated bruising of the hide could impact on its long-term durability. It did, however, have the effect of preserving the skin in two ways.

First, the fibres were separated out and then saturated in fats to the point where they were insoluble. Second – and perhaps more importantly – through a process of oxidation, brought about by the inclusion of fatty acids, the hide was further preserved against putrefying. Oil tanning didn't die out completely with the introduction of more modern methods. Well into the late medieval period oil-tanned hide was in popular demand for protective garments, and even today fish oils are used to create a suede-like chamois leather on an industrial scale.

But over time another process evolved that rendered the pelt permanently imputrescible while also preserving intact the natural fibrous structure from which its strength and pliability are derived. Strabo, the Greek geographer, described leather as one of Britain's chief exports, and Julius Caesar in his written account of the Gallic Wars distinguished *aluta*, the soft leather used in sails, from *corium*, the hard leather used by the British to line and waterproof the outsides of their boats. So it's likely that the British, as elsewhere, had developed the technique of tanning leather by at least the first century AD. A number of factors may have influenced this successful export trade. In the early part of the first millennium AD, Britain would have had many more acres of oak woodlands than it has today, and both oak bark and galls (oak 'apples') are key ingredients in the tanning process. British leather was also of such high quality because of the particular breeds of cattle. Any industry is reliant on its supply lines, and the export of British leather was only as successful as the production line that served it. Understandably, the heaviest and thickest hides came from the beasts raised in the most extreme environments, where their skins had to deal with the wet, wind and cold for sustained periods. The mountains of Scotland and Wales, the hill ranges of northern England and the rugged moors of the south-west have long supported livestock industries that even today are famous for their breeds of cattle.

The tanning of hides to produce leather of the highest quality is one of the true and original skills of human craft. It's a knowledge wherein precision machinery, chemicals, engineers, dials and gauges, formulae and test tubes have entirely replaced a series of manual operations that demanded a high level of *cræft*. It's a far from simple process, taking around eighteen months from start to finish, and one that involves an enormous variety of techniques that can at any point go wrong, with ruinous consequences. To begin with, if beasts were slaughtered on-site, the skins could be taken directly to the tanner for processing. Being such a lengthy process, a fixed, spacious and stable industrial base was required, and one with a fresh water supply to conduct rinsing operations.

The hide itself has three layers. The first, the epidermis, is the breathable, and thus perishable, membrane that contains the hair follicles. Beneath this is the corium, the true skin consisting of a felt-like mass of fibres. The adipose is the connective tissue between the skin and the flesh, and this, along with the epidermis, must be first removed in the tanning process. To achieve this, the hides were soaked and washed to remove any albuminous material – water-soluble filth such as dung or earth that may have bonded to the fur. Sweating, where the hides are folded into themselves with a top dressing of urine, was used to break down this first layer.

A speedier and more industrial method would be to immerse the hides in a solution of lime, the caustic material produced when limestone is roasted at temperatures of over 1,200 degrees centigrade. A slaked lime solution weakens the follicles and speeds up the process by which the epidermis would naturally deteriorate. The tanner and his team would then set about manually stripping the hide of the epidermis and the adipose layers. This they would do on a 'beam' using a technique known as 'scudding'. It's this part of the process that is the most iconic in traditional tanning, an image of the tanner bent double over the sloping beam, working up and down over the hide with a blunt draw bar scudding the decaying epidermis and hair away from the corium. A sharper blade would be used to achieve the fleshing on the flip side as the remaining adipose tissue was removed. Further episodes of liming could be conducted at this point but, to improve suppleness in the end product, the hide could also be submersed in a solution of water, bird droppings and dog faeces. This would help drive out any remaining lime solution and had the additional benefit of plumping the hides, swelling them so

that they would be perfectly prepared for soaking in the tanning
solution in the next and definitive process.

TANNINS OCCUR NATURALLY in some plant materials,
and in Britain by far the most abundant source is found in
the bark of oak. Chemical and mineral tannage is used today,
and the traditional use of natural plant materials has therefore
come to be known as vegetable tanning. But it's not until this
process has been conducted that a hide can be considered truly
imputrescible or water-resistant. The tanning process involves a
chemical reaction whereby deposits of very fine sediment are left
in the interfibrillar space of the hides. This 'bloom' of ellagic acid
and its antiproliferative qualities inhibits cell growth and thus
the process of bacterial decomposition. But this is a process that
takes a considerable amount of time.

In some late-medieval guild statutes it is considered criminal
to subject hides to anything less than a full twelve months of
submersion in an oak-bark solution, with an eighteen-month spell
favoured for the thickest hides. Various methods can be used to
speed up proceedings, but these are only conducted to improve
the quality of the tannage rather than to cheat the customer.
Regular paddling, sorting and moving of the hides is one way
to increase penetration, but this should be done in the gentlest
of fashions in order that the hides are not bruised and damaged.
In mechanised factories drums and rockers do this work, but
traditionally it was conducted by 'handlers', whose job it was to
move hides at various stages through a sequence of tanning pits
in order to achieve the best and most even tannage.

In the tanning pits was an infusion created by leaching oak
bark in water. There were a number of pits ranging from those

containing the freshest infusion to those that had already been in service for some time. The older the solution the mellower it would be, and it was into these weaker solutions that the new or 'green' hides were introduced. The danger of introducing green hides into a newly prepared infusion of oak bark is that the solution works too quickly on the outside of the hide and fails to penetrate into the heart. Better to start the process off slowly and work up to a greater strength of infusion. Thus, as hides moved up the sequence of pits towards the freshest and most potent infusions, the infusions themselves would move down the sequence towards the greenest hides. In this way the most effective and even tannage would be achieved across all the hides in the tannery. In the very late stages, it might be that the hides were stacked in a pit on top of each other with ground tannins sprinkled between the layers. This concentrated exposure would help to finish off the hides ready for final extraction and soaking before drying, which was undertaken as slowly as possible and never in direct sunlight.

FROM THE TANNERY, leather hides would be sold into a range of industries for further processing. A cursory list of traditional leather products might include gloves, breeches, braces, belts, hats, hoods, aprons, armour, jerkins, neck pieces, gauntlets, boots, gaiters, chaps, helmets, water bottles, bellows, drinking vessels, jugs, travelling packs (for people and horses), buckets (particularly for gun powder), ink wells, water or wine carriers, book binding, chests, tents, wallets, pouches, sails, boat lining and harnessing. Far from becoming obsolete with the onset of mechanisation, leather was to prove invaluable for producing drive belts for engines; hoses, valves and washers for pumps;

hoods for the roofing of carriages and coaches; and upholstery for home furniture. It has left its mark on sports and recreation too, for without leather we would be without cricket, rugby and football – three sports that represent some of our biggest global exports.

What is perhaps most fascinating about the production of leather is that not only is it a by-product of the meat industry but, with the exception of the lime, every material used in its production is either a by-product of another industry or domestic waste. Whether urine, dog faeces, bird droppings or oak bark, these key ingredients in the traditional production of one of mankind's most versatile products could all be sourced locally and at the cost only of transportation and handling. It may not seem that chucking some skins into a few pits filled with the foulest and most noxious solutions represents a particularly knowledgeable and hard-earned set of skills, but the resourcefulness of tanning, the expert methods of its processing, along with the multiple uses of its products, represents a *cræft* of almost unparalleled importance to the human story. To come back to Ælfric's *Colloquy*, and the final line with which the shoe-wright justified his craft, the critical importance of leather can be gauged by the most basic of functions that it has fulfilled in the narrative of human development. The astute observation that nobody could wish to go through the winter without *minon cræfte* is in many ways an understatement of exactly what leather has allowed us to do as a species.

Leather has allowed us to survive the harshness of the cold and penetration of the wet and to advance our positions as we undertake the seasonal work of the winter months: the felling in the woodlands, the draining on the marshes, the dunging of the fields and the hedging of the paddocks. It has allowed us to stand fast in the face of the worst that the colder climes can throw at us and to come away as winners. But it has also allowed us, in a world before machines, to *be* the machine. Where now the mechanical flayer cuts back the brash of the hedgerow and clears the path, gaitered, jerkined and gauntleted workers once toiled. Of those who cut, sliced and hammered in industries so dangerous that now only a machine is considered safe for purpose, where would they have been without the leather apron and leather-gloved hand? And harnessing may seem like a throwaway triviality in the minds of modern readers, but the reality of effectively ensnaring the power of the ox and the stamina of the horse and turning it to our advantage, in the field and behind the cart, has afforded us a capacity on which cities, states and empires have been built.

And finally, one of the greatest allowances of leather is the manner in which it has facilitated travel. In an age when to travel was to communicate, leather gifted us the ability to move through territories of varyingly extreme conditions, to accelerate interactions, to advance ideas and to discover. Through the saddle and the reins we have carried works of wisdom bound in leather. We have ridden into snowstorms that have beaten against but not broken our leather cloaks. We have crossed arid deserts safe in the knowledge that the water in our flasks would see us to the next oasis. In short, of all the materials at our disposal, it's one of the oldest that remains one of the most desirous. And though a profusion of alternative and imitative products are available today, few would accept any other material to clothe their feet. Even now, nobody would ever dream of attempting to get through the winter without the shoe-wright's craft.

M Y PASSION FOR leather is most evident in my collection of shoes. It all began with my grandfather on my mother's side. He'd been at the first Siege of Tobruk in 1941, when the Axis forces had surrounded the Libyan Mediterranean port as the British Western Desert Force regrouped. While unloading reinforcements and precious supplies, he'd found himself out in the open during a dive-bomb offensive. A squadron of the Luftwaffe's Stuka bombers screamed down on him and his fellow combatants, and as a result of the attack a piece of shrapnel flew through my grandfather's lower right leg, shattering the fibula. In a long and morphine-induced daze, Private Alfred Kenneth Collis was shipped back to Blighty, a journey during which he vehemently resisted amputation.

He wasn't a vain man but, in his words, he didn't want to be a cripple. As a consequence he suffered terrible episodes of infection in the wound for a twenty-year period that often saw him hospital bound. In the end, he was right to be so insistent during that journey home. A new surgeon – an 'Asian' as my grandfather was always proud to say – came along and proposed removing all the damaged bone, leaving the tibia and surrounding muscles intact. And, although the result was that one of Grandad's legs ended up an inch and a half shorter than the other, it worked. To make up for the discrepancy, he was directed to the National Health Service's orthopaedic department to have some special shoes made for him. As I said, he wasn't vain but he didn't want to wear shoes that made him look like a cripple. These shoes came to be known, by the war-wounded soldiers who had to wear them, as 'elephant boots'. So, instead, he went to a high-class shoe shop, bought a number of pairs of handmade brogues and had the heel on the right shoe built up by a local cobbler.

Grandad was my hero. He'd fought against the Nazis and had the scars and the limp to prove it. Perhaps the most heroic thing about him, though, is that right up until his dying days, he never talked about those quayside events of that fateful day in 1941. As kids, we would beg him to tell us what had happened to give him his limp, because we genuinely wanted to know but also because each time we asked he'd come up with an even more amusing tall tale. My personal favourite was that he was leaving the cinema late one night and as he exited the building he saw three gorgeous young ladies on the other side of the road. So enraptured was he by their beauty that, not looking where he was going, he tripped over the kerb and broke his leg. After a while, we learned not to ask.

I also learned to love my grandad's brogues. I never thought I'd find myself wearing shoes like that – they seemed so unfashionable to me as a kid. Nonetheless, I eventually warmed to their shape and form, probably because they'd helped my grandad maintain a shred of normality and dignity in post-war London life. It wasn't until I was in my early twenties that I purchased a pair. In truth, I had little in the way of options. I can't remember exactly where I was at the time, but I was jobless and had decided to go walking the ridgeways of the Cotswold Hills like a vagabond. I was traipsing from village to village with just a camera, a notebook, a map and a blanket, when on one particularly rocky stretch of hillside the soles on my trainers gave up and fell off. I was truly down at heel, penniless and barefooted. But the sun was shining and the air was clean, so I hobbled down into the local town, where I was lucky enough to find a charity shop. And the only shoes that fitted were a pair of tan Richleigh brogues.

Now, I don't know how Cinderella felt when she first slid her delicate feet into those glass slippers, but for me it was a life-changing moment of fairytale proportions. I wore those brogues everywhere. I walked in them, dug in them, partied in them, went to meetings in them, and wore them in the sea, around the farmyard and on the boat. I polished them religiously to keep them in good shape, and when the soles were worn through I simply took them to the cobbler and had a new set stitched on.

They're in retirement now, stowed in a shoe cabinet of brogues, dealer boots, dress shoes, loafers, Chelsea boots and other renowned types of leather footwear. But if I had to choose just one pair of shoes to get by with, it would almost certainly be those tanned brogues. The fact is, they are a remarkably versatile shoe – the high point of a coming together of a simple craft and a wonder material.

IF THERE IS one other craft where the role of leather has caused me to marvel, it must be the harnessing of the power of draught animals. It's said that while a man might be persuaded to wear a badly fitted suit, the horse can be surprisingly unco-operative if not fitted with the correct attire. But get the harnessing right, match it to the form and strength of your horse, and it will be a willing and able partner in pretty much any endeavour. Our relationship with horses, from the earliest days of domestication through to the present, could not have achieved what it has without the manufacture and crafting of leather. Obviously, as both steeds and pack animals, they have allowed mankind to move at greater speed and with a greater volume of supplies. They have drawn carts and wagons, which has increased our capacity to transport and therefore to expand our towns and cities. But utilising their strength and intelligence has been of most value in the ploughing of fields, the turning of the sward and the production of food.

A few years back I attended the Southern Counties Heavy Horse Association's annual All England Ploughing Match, one of the most popular, long-running and prestigious matches on the circuit of regional horse-ploughing competitions. This event draws a large contingent of ploughmen and women from all over the British Isles, and for that year's feast of furrow-turning the Association had secured an ideal showground. It sat alongside a main trunk road, with a soil light under the foot, a south-facing aspect and the perfect arrangement of fields to accommodate the plots, craft and food stalls, competitor stable-boxes and extensive car park. Organisers, competitors and exhibitors alike were hopeful that, after two years of complete washouts, this year the event would experience something of the glory it had displayed back in its 1950s heyday. But as the weekend drew closer the forecast was looking ever more ominous. On the eve of the event I resigned myself to fate. I packed some industrial footgear, a heavy raincoat and my waxed trousers.

On arrival at the site the next morning my heart sank. But not as deeply as the horseboxes had sunk into a thickening morass of mud at the site entrance. The tractor belonging to an obliging local farmer rescued the most desperate from the mud while others unloaded their horses and cargo where they'd been grounded then made their way over to the match plots. Stewards directed the public to the drier parts of the car park as exhibitors and stallholders wrestled with ferocious winds in a bid to get their marquees and gazebos up in time. By ten o'clock, the beer tents, burger vans and fish and chip bars were in place. A modest flow of spectators braved the elements and as the first teams took to their plots, I was determined to ignore the wet and cold and enjoy my day. It was hard, though. In less than an hour a light shower had developed into a sustained downpour. A brief flurry of activity as spectators pulled on their waterproofs and raised their umbrellas was quickly followed by an almost complete abandonment of the showground. The forecasters had been right and an onslaught of south-westerly squalls set in for the afternoon. By two o'clock, all but the most dedicated spectators – maybe four or five people – had left, and stallholders were packing up, slipping and sliding in the mud as they folded canvas and ferried equipment to their stranded vehicles. As a public spectacle, the All England Ploughing Match was effectively over.

Nonetheless, I was beginning to enjoy myself. There's nothing better than adversity to summon up the blood. And when I realised that not a single contestant had the slightest inclination of giving up, I was further spirited. Despite the atrocious conditions, with soil clagging boots and hooves and clinging to the hard steel of the ploughshare, the standard of ploughing was impeccable. Anger, frustration and exasperation were met in equal measure with banter, laughter and camaraderie. Up and down they trudged all afternoon, their bent frames curled against the lashing of the rainstorm. Grudgingly, the horses obeyed increasingly desperate commands as the contestants meticulously tweaked their ploughs' settings, making microscopic adjustments to the ploughing width and depth. While finishing was important, winning was everything. These ploughmen and women were purists, perfectionists and battlers.

As the afternoon wore on, almost as a reward the sun sank below the storm clouds and bathed the scene in a majestic pink twilight. Rain glistened on furrows, steam rose from horses' backs and a brilliant light danced off the polished brass harness fittings. By now, most of the plots had been completed with only a couple of teams still in the field. As I gazed out across the beautifully crafted furrow slices and watched the final team plough their last run, I was struck by the scene's familiarity. It was one I'd seen on numerous occasions as a boy. Not for real but depicted in a tobacco-stained print on the wall of the local pub near my childhood home in Sussex. I used to gaze up at this cliché of yesteryear, representing the immemorial relationship of man, beast and landscape. The ploughman's collar was drawn up tight against his jowls, the horses strained into the harness, their manes curling against the rain, and the landscape revealing farmstead and labourers' cottages nestled snugly in the valley below. There was an intense solitude in the ploughman's work, a relentlessness to the task at hand and a pathos that tempered even the most ardent boyish romanticism. And now here I was, in a sodden Wiltshire field thirty years on, seeing for real the image that had beguiled me as a child. Only this wasn't for real. It was the tail end of a great tradition, the last vestige of a dying way of life.

The piercing crackle of a public address system startled me from my reminiscence as the chief steward beckoned the competitors forward for a makeshift award ceremony. A ragtag bunch of muddied and hooded competitors trudged towards a solitary caravan, where a party of quasi-official delegates, myself among them, handed out rosettes and silverware. Later, as I drove home, I wondered what had gone wrong. Okay, the British weather is famous for ruining even the most prestigious of outdoor events, but why had an occupation so ingrained in our rural heritage found itself clinging so desperately for survival as a public spectacle?

We have a strange relationship with the past. We spend millions on monuments and memorials celebrating the legacy of elite national heroes, yet when it comes to a craft so intimately bound up with our landscape story and rural ancestry only the dogged enthusiasm of a few dedicated fanatics keeps it alive. Was ploughing with horses a practice destined to be depicted as it was on that faded print in the country pub – an outdated and ultimately redundant activity, nothing more than romantic musing? Using draught beasts to work the ground to produce food is a practice with a history of around eight thousand years, and a story that is intricately linked with that of human evolution. In this country the earliest evidence for ploughing with draught animals comes from the late Neolithic (3000–2500 BC), and I can recall the intense fascination when I'd helped excavate a Bronze

Age barrow and found evidence of ard marks, a primitive means of ploughing, on the ground surface beneath it. So were the events of this disastrous October afternoon all that were left of this rich cultural tradition that had served humankind so well for so long? Should it not be better recognised? And what are the dangers of losing these skills altogether?

THINK OF THE countryside today and it's difficult to imagine it without the omnipresent tractor ceaselessly working its way up and down the fields. And yet the tractor is a relative newcomer. Prior to the mechanised horizon of early twentieth-century British agriculture, it had been farm labourers and draught beasts that had borne the brunt of the hard work. This made them the last in a long story that has its beginnings in the fertile crescent of Mesopotamia, which broadly speaking is the land around the river basins of the Tigris and Euphrates. While the earliest depictions of oxen being used to drag an ard – consisting of little more than an angular stone lashed to the end of a stick – come from the Upper Nile of the fourth century BC, archaeological evidence suggests that oxen had been domesticated to work the land as early as the seventh century BC. Oxen then continued as the ideal draught animal for the next eight thousand years, with donkeys, mules and in some places goats used to a lesser extent.

Knowing exactly when horses started to replace oxen, especially in Britain, is difficult in the absence of solid evidence. The traditional view was that the Normans were largely responsible for the wider introduction of the horse to England in the late eleventh century (crudely deduced from the Bayeux Tapestry), but it's now clear from numerous documentary references, as well as archaeological evidence (especially from richly furnished horse burials), that horses formed a significant part of the day-to-day workings of Anglo-Saxon society. But whether they were used for draught purposes is a matter for conjecture. A horse is depicted drawing a spike harrow in the lower border of the Bayeux Tapestry (1077), and William Fitzstephen informed us, in 1174, that horses were being sold 'for the plough' at Smithfield Market. But even when other parts of Europe had successfully made the switch from oxen to horses (certainly by the fifteenth century in France and the Low Countries), in Britain oxen remained popular. One interpretation sees this as a result of the Englishman's predilection for roast beef. After all, a breeding strategy aimed at producing generation after generation of working oxen would almost certainly have led to more cows in the countryside. And oxen continued to play a role well into the twentieth century, particularly on heavy clay land where their slow and steady pace enabled them to work soils on which the more spritely horse would have struggled. Compare a photograph of oxen ploughing on the South Downs of Sussex in 1905 with the image of oxen ploughing in the Harley Psalter, an illuminated manuscript from the first part of the eleventh century. There is little difference.

By the seventeenth century it seems that horses were doing a large part of British farm work, a development which may have come about as a result of changes in warfare and a shift from the heavy cavalry associated with medieval field combat to the light cavaliers made famous in the English Civil War. The latter required much lighter and swifter steeds, and it's possible that an abundance of heavy horses, and the long-standing and prestigious culture associated with their breeding, were transferred to farmers from their gentry landlords in a desire not to allow a great tradition of heavy-horse breeding to die out. From the mid-eighteenth century we can track the development of the horse in agriculture as studbooks, sales ledgers and the records from county shows began to appear in published form. Equally, the agricultural implements from these periods reveal the changing nature of cultivation and improvements in harnessing. While the selective breeding of horses had always been a consideration for farmers and landowners, it's in this age of agricultural progress that we see the adoption of quasi-scientific models.

New breeding principles, aimed at developing certain breeds of sheep and cow to better suit them to their environment, were now also applied to horses and, by the end of the nineteenth century, resulted in three significant breeds of heavy horse: the Shire, the Clydesdale and the Suffolk Punch. The Shire can be seen as the English heavy horse, originating in the central shires of England, whereas the Clydesdale, with its origins in the band of lowland between Glasgow and Edinburgh, is the Scottish representative, and the Suffolk Punch was the horse of choice in East Anglia. Of course, there were numerous other breeds, such as the famous Welsh Cob and the Exmoor Pony, that were better suited to the rockier terrain and cultivation practices of their respective heartlands. But the main objective in the selective breeding of the heavy horses, apart from their docile and obliging nature, was their ability to pull more and for longer – brute strength was everything. Thus, at the turn of the twentieth century, the practice of using animals to work the land to produce food had reached a high point of technological innovation.

It's quite some journey, from the seventh century BC scratch plough once pulled by a single yoked ox over light sandy soils to a harnessed team of four or more horses capable of pulling four steel ploughshares through thick, heavy clay. As civilisations and empires have come and gone, the method of turning the soil to prepare a seedbed has continued to develop and be subjected to innovation. And what is most intriguing is that while breeds and species of animal have changed and the types of plough have been incrementally improved, one thing has stayed the same: the leather in the harness. Despite all the technological developments of the past 250 years, there is not a single material that can compete with leather for strength, durability, comfort, breathability, flexibility and availability. Like those Richleigh brogues that moulded to the shape of my feet, the harness of a draught animal – the collar, bridle, breeching, girth straps and ridge pad – within a period of only a few workings will curve and form around the beast to best channel its power.

Does it serve us well to put aside centuries, if not millennia, of technological development and breeding innovation in favour of the more powerful tractor? Might there come a time when the tractor's vastly higher levels of productivity fail to offset favourably against its outlay and running costs? If so, might humankind once again consider the virtues of the horse as a source of farmyard power? Speculative questions, yes, but while horse power is seen as an indicator of backwardness among the rural poor of Eastern Europe, in some parts of the US certain groups – chief among them the Amish – successfully demonstrate that horse power falls not so far behind the efficacy of fossil-fuel generated power. Moreover, the Amish people are experiencing a population boom across the US and Canada, and their success is in no small part down to their attitudes towards community and food production, in both of which the horse plays a prominent role.

Fascinatingly, when horse draught power was first confronted with a competing engine power in the mid-nineteenth century it shrugged off the challenge. In the early decades of Queen Victoria's reign, steam power was heralded as the only energy source that could satisfy the increasing demands of a fast-industrialising nation. In virtually every industry, steam machines came to play a fundamental role, and it was off the back of this innovative technology that Britain rose to prominence as a global superpower. The agricultural economy also found itself benefiting from the introduction of steam machines where water pumps, threshing boxes and milling equipment could all be run much more productively from this hugely powerful source of energy.

But the one area where steam failed to make its mark was in draught and cart work around the farm. Although steam ploughing did experience a short bout of enthusiastic promotion in the mid-1800s, it was swiftly realised that the horse remained the more effective and productive means of growing food on the farm. More versatile, more nimble and with fewer energy demands, the horse demonstrated to even the most forward-thinking farmers that it had yet to be surpassed. Crucially, the main benefit was that the horse was light on the land. Farmers keen to try out the latest in steam technology soon discovered that driving twenty tonnes of engine over their land would quickly compress the soil. As a consequence, their fields would develop what are known as 'pans', buried, impenetrably compacted layers that prevent moisture and minerals from passing between the topsoil and subsoil. Panning can have a disastrous effect on crop yields, and alleviating it required extra phases of deep ploughing and subsoiling, which in turn required ever more energy expenditure. Of course, the same rules apply today: the use of heavier tractors commits the farmer to a regular programme of deep ploughing. Not only does regular subsoiling, which is needed to break up these pans, require extra fuel but it damages soil depth and quality by destabilising it, making it prone to events of rainwash and disrupting the micro-environments in different layers which would otherwise help to break down organic matter into nutrients.

I see the use of leather for the harnessing of horse power for food production as one of the *cræft*iest of all crafts – but especially when it's compared to the tractor. What makes the horse so attractive is the fact that all its fuel needs can be provided by the farm itself. Horses need little more than grass to graze on and, having fattened on lush pastures over the warm summer months, they will be happy enough with hay throughout the winter. But if you're going to work them hard they require a little extra in the diet, most commonly milled oats, beans or a barley-meal, which gives them the protein and carbohydrate hit they need to sustain long hours working in the wet and cold. All these sources of fuel can be grown on the farm, and a shift to organic forms of fertilising – such as a return to the traditional inclusion of nitrogen-fixing clover in crop rotations for growing cereals – would improve the soil and provide a voluminous quantity of fodder for the horses.

It might be argued that a tractor could contribute to the production of its own source of fuel in the form of biodiesel, but this is where the horse really outstrips the tractor. For even if the tractor did help produce its own fuel, such as crops of rapeseed or sunflowers, it would still require an on-site processing plant – a reactor, a settler, a purifier and an evaporator, all of which require fuel – to convert the raw biomass into fuel. The horse, on the other hand, needs only its teeth, stomach and lower intestine to convert a bellyful of oats into a day's work. The horse also produces a by-product of dung, and the payback here is that its own manure can be ploughed into the fields to further improve fertility and productivity.

If the fossil fuels on which all our food production is reliant rise in price, then it isn't only their costs that matter. The tractor is itself a product of numerous fossil-fuel dependent industries, and its production would therefore see a concomitant rise in the entrenched energy costs of manufacture. The primary industries alone (mining for iron, copper, lithium, aluminium, and all the other many elements that go into its production) consume vast quantities of fossil fuel before manufacture and assembly take place. Moreover, tractors already represent one of the biggest capital investments a farmer has to make, with some models costing over a quarter of a million pounds. Any rise in this cost would have to be transferred to the price of the food. What is perhaps most scandalous about tractor production – and it's the same for most manufacturing industries – is the level of planned obsolescence embedded in their engineering. This built-in redundancy means that on most western farms a tractor will be sold on to the developing world before it reaches the end of its working life. Put simply, it's cheaper to buy a new model than it is to maintain and repair an old one, and as a consequence, on most commercial British farms, tractors don't see use beyond their fifteenth birthday. So not only is a horse capable of reproducing itself, when worked between the ages of three and twenty-five it has a longer working lifespan than a tractor, and it improves with age, peaking in its mid to late teens.

In all this, the one issue that cannot be ignored is people. Working the land intensively with horses would need an army of ploughmen, farriers and stablehands. To repopulate British farms in this way would require reversing the two-hundred-year-old trend of people fleeing the relative poverty of the countryside for the opportunities and wealth of the towns and cities, but with the rising price of food it may soon be cheaper to pay people rather than paying for diesel to grow our food. The financial incentives may one day arise. But there already exists a vast body of people who spend as much of their free time as possible with horses. A recent survey by the British Horse Industry Confederation found that there are over 1.3 million horses in Britain today, with an estimated consumer spend on their upkeep and care of around four billion pounds per year. Horses have never been so popular. While the overwhelming majority are dedicated to recreational pursuits, there is a vibrant skill set out there and a continuing affinity with and affection for the horse. Perhaps this is a result of thousands of years of symbiosis. And just as we have caused the horse to evolve into the many breeds that exist today, so we have evolved to admire and respect it. We often talk of the value of a 'people person' – the ability to communicate with others and improve the way they do business. But equally there are 'horse people', who are often happiest when working quietly, assuredly and thoughtfully with their animals. A reservoir of natural aptitude is out there ready to be tapped into.

Finally, there is also the critical issue of our dependence on oil. Subject to the uncertainties of the free market and the effects of geopolitics on fuel prices, producing food with draught animals creates a certain degree of resilience. No matter what happens over the hill, with a horse and plough you'd still be able to feed yourself. And this may be why mankind has plodded on for millennium after millennium, reaching an ever more developed state of material complexity. Humans have survived because of an ancient practice intrinsically based in the local – a fundamental connection to the land on which we live. Having lost that connection we now find ourselves at the mercy of the global winds of change, relying almost religiously on the vision of a hi-tech future to solve the crisis of ever-rising food prices and population growth. The horse will undoubtedly take some beating as a solution to low-impact, locally sourced food-producing power. Even better, we have inherited centuries of selective breeding that has produced horses for courses. The hard work has already been done for us. We don't have to embark on a fact-finding, in-development phase of transition to test and refine our machines, for we can just harness our horses and pick up where we left off a hundred years ago.

Of course, we might have to ask the ragtag bunch of ploughmen who braved that October deluge at the All England Ploughing Match if they would be kind enough to harness their draught horses and show us how to take up the reins . . .

10

SEED AND SWARD

Hwilc þe geþuht betwux woruldcræftas heoldan ealdordom?
Eorþtilþ, forþam se yrþling us ealle fett.

Ælfric's Colloquy

[Master] Which do you think, amongst the world-crafts, holds
the most authority?
[Pupil] Agriculture, for the farmer feeds us all.

EVER SINCE OUR first meeting, Janet Mudge had insisted that my wife and I should come up to their farm for one of her special Sunday roast dinners to experience what she called, 'the true taste of Devon'. Reared on the farm, slaughtered locally, hung for four weeks and slow-roasted on their farmhouse kitchen stove, the Mudges' Sunday roast has remained a culinary legend in our household. You will struggle to find beef that tastes quite like it.

At the time of my first meeting with Janet and her husband, Francis, I was making the BBC series *Edwardian Farm*. The idea was for a small team to run a modest farm set on the banks of the River Tamar as authentically as it would have been in the opening decade of the twentieth century. *Edwardian Farm* was a follow-up to a previous series, *Victorian Farm,* during the making of which we had relied heavily on the authority and practical knowledge of a certain Thomas Stackhouse Acton, Esquire. Tom Acton, whose estate in Shropshire has been in the family since the thirteenth century, had become an agricultural mentor to me as I embarked on a year as a Victorian farmer. In Devon, for *Edwardian Farm*, we'd struggled to find a similar character: an on-screen role as our farming expert but also someone from whom I could seek advice on some of the finer details of growing crops and raising livestock in this part of the country. I travelled far and wide to small farms throughout the province in search of just such a person, and I did come across many extremely friendly, knowledgeable and forthcoming individuals who went on to contribute, whether on or off screen, in a number of our programmes. But none seemed to be cast quite as neatly into the mould of the farming expert we all wanted.

I was at the point of giving up when I stumbled across Francis

Mudge. It was late winter and as part of our farming year in the south-west we felt it imperative that we cover the story of the market garden industries of the Tamar Valley, which had reached a high point of production in the Edwardian period. I was charged, single-handedly, with recreating a market garden – no mean feat – and spent the vast majority of my spare time clearing weeds from some south-facing slopes near the farm and planting out young strawberry plants.

The time came in early spring when the white flowers signalled the coming of fruit, with each flower representing a potentially succulent and profitable strawberry later in the year. But they are profitable only if you lift them off the ground to protect the emerging fruit from the damp, the mud and the greatest enemy of all, the slugs and snails. One traditional method of doing this – still practised by the small-scale organic growers today – is to use straw as a bedding and mulching layer. Not only does a thick carpet of straw protect the fruit but it can also protect the soil from the sun's heat, trapping moisture in the ground and allowing the fruit to swell to larger and juicier proportions. The problems I was having in sourcing straw for this purpose were both historical and practical. In the first instance, I found it difficult to lay my hands on straw at a reasonable price, for Devon is not especially renowned for cereal growing (from which straw bales are a by-product) and what little was available on the market was expensive owing to the demand for vast quantities of bedding material for lambing at that time of year. Similarly, the sheer number of market gardens along the banks of the Tamar in Edwardian times meant that straw would also have been in relatively high demand just over a hundred years ago. Growers back then had to resort to other methods, and in reading some of the accounts of these early horticulturalists I came across what I thought would be the perfect solution for replacement straw: my old friend, bracken.

As I'd found through my thatching experiences, there is never a bracken shortage on the acidic soils in the west of England and Wales. The problem I had was that, despite its prevalence throughout the surrounding landscape, I needed a way of collecting it up and delivering it to my strawberry plants. I'd thatched a barn with the dreaded stuff and I was damned if I was going to personally source enough to line a whole hillside. This was when Francis Mudge stepped into the breach. A local woman told me about a farmer, up on the edge of Dartmoor, who cut and baled bracken as a means of fighting back against its rapacious spreading and as a way of making a few pounds by selling it locally as a bedding material. I acquired a telephone number and called to make some enquiries. I spoke to Janet – Francis had a hearing aid and never answered the phone, so I ended up doing all my business through Janet – and arranged for eight bales to be delivered, as they say in Devon, 'directly'. Pronounced 'dreckly', this word is the Devonian equivalent of the Spanish *mañana*, meaning, 'at some point in the near future but exactly when is anybody's guess'. Later in the week I was at the farmyard cleaning out the cattle stalls when a call came up from the cottage that a Mr Mudge had arrived. So we arranged to meet him in the lane, fired up the Land Rover and set off.

The moment I clapped eyes on Mr Mudge it occurred to me that we might very well have found our on-screen farming guru. He certainly looked the part, with his mutton chop sideburns and the wind and weather of at least fifty years of farming ingrained in his face. And he talked the talk. No sooner had we shaken hands and looked over the bales of bracken than he launched into a string of questions about our farming endeavours. We parried with our own questions about what we were getting wrong and – crucially – about the old ways of doing certain things. This is where Mr Mudge really came into his own: he had a surprising awareness of the varying approaches that could be taken to particular farming processes depending on the technology one had available, and, along with his understanding of pre-war methods, this made him a priceless contributor. What I really liked, though, was his enthusiasm and the fact that he'd tried many of the old methods himself, just for interest's sake.

We unloaded the bracken bales then we all marched up to the farm, where Mr Mudge proceeded to wax lyrical in his quiet but authoritative manner on all the things we needed to do and why – citing examples from his own experiences as well as those from other farmers he'd known over the last half century. As he cast his eye over the field where I was hoping to sow a bag of oats he detailed exactly the method of improving fertility that I'd read about in the historical farming journals of the time. I grinned and thought to myself, 'We've found our man.' Not just somebody to play the required role on camera, but an invaluable source of local knowledge. As our first afternoon together drew to a close, we'd pretty much exhausted most topics of conversation. Janet Mudge, who had been with us the whole time, made up the other half of what was a bit of a double act, prompting Francis and drawing on details from her own memory archive when the salient point of a good yarn escaped him. Off the back of a discussion about what we needed to do with our modest herd of cattle in the forthcoming months came Janet's offer of a roast dinner.

Before our year in Devon came to an end, we headed over to the edge of Dartmoor one Sunday afternoon. In truth, it wasn't just the promise of a roast that lured me to this ramshackle farmstead clinging tenaciously to the foothills of one of England's last great wildernesses. There was something else. Something that up until about two weeks earlier I hadn't even known existed but which, for me, held the key to the most traditional of farming practices in the south-west. It was during one of Mr

Mudge's frequent visits to our farm that we'd got talking more generally about the style of farming in Devon and Cornwall and how it differed from what the locals called 'up-country'. In this conversation he alluded to a particular contraption that he'd only just managed to lay his hands on and had yet to see working. I was beginning to realise why Mr Mudge's experiences were proving so invaluable to me, for in my eyes he was of a dying breed – a type of farmer who managed to eke out a living from a relatively small farm set on particularly poor soil. To make ends meet he'd been forced to diversify, something that, in an age before national and global transport networks, most British farms – even those on far better soils – had had to do out of necessity. As a result of that diversification, all resources on the farm were valued and nurtured for the contribution they could make to the overall worth of the holding.

Today, most farms specialise in a way that results in many areas of the farm being neglected in favour of the production of only a handful of commodities. This can be dangerous as it means putting all your eggs in one basket (quite literally, if you specialise in poultry), but when you get it right it can reap dividends. With diversification you might not get quite the same cash crop at the end of the year, but if one line of production falters, say, because of bad weather or disease, there are always other parts of the farm to fall back on, and a steady trickle of income rather than an annual bumper harvest is what keeps the wolf from the door. This is what I call *cræfty* farming. The main issue is that diversification requires a greater skill set and knowledge base, not just of the many and varied processes it takes to make a wide variety of commodities financially viable but, more critically, in their successful integration. And this was second nature to Francis and Janet Mudge.

At about this time I'd been reading a report on environmental core samples dated to the eighth century, which had been taken from wetland areas on the fringes of Exmoor. The wet conditions meant that the preservation of organic matter in these areas was excellent, allowing environmental archaeologists to identify a range of seed and pollen types. This in turn enabled them to reconstruct the farming practices of the area in the period from around AD 600 to 800. The findings were revolutionising and challenged our understanding of early medieval agriculture. For example, it had traditionally been believed that the iconic nucleated village and its associated medieval three-field system originating in the period from around AD 800 to 1100 represented the most effective and progressive form of agriculture of its time. This was a system of three vast open fields and each year each field was divided into strips and shared out among the village community for them to work the land 'in common'. While fields one and two were sown down to crops, the third would be allowed to lie fallow, to be grazed by the village livestock and rested for a year.

The area the three-fields and their concomitant nucleated villages came to dominate became known as the 'central belt' of England, cutting a vast swathe from the north-east down to the south country, at its most pronounced in the East Midlands. It appeared to have been imposed on an earlier arrangement in the landscape, and marked an apparent shift from individual holdings to an arrangement where everyone appeared to work together. This led many to see the three-field system as innovative and more productive. By implication, the farming practices of the regions to the east and particularly the west of it were seen as backward, and as having neither the political structure nor cultural adaptability to grasp these new ways of working the land and of organising people.

But the environmental core samples from areas around Exmoor were beginning to change this view. What the pollen sequences suggested was that a complex system of farming had indeed existed in these areas, one that was surprisingly versatile and enabled farmers to flexibly alternate between cropping and livestocking – between seed and sward. A remarkable seven-course rotation seems to have promoted cycles of wheat, barley and legumes interspersed with years of grassland pasture. In this system, if the fertility of the soil was in question, perhaps because of poor yields, a field could be reverted to grassland and grazed by livestock, which in turn would replace some of the lost fertility with their droppings. So this worked in much the same way that the fallow year worked in the three-field system. Yet, because this part of the country had retained its arrangement of small fields and dispersed settlements, the farmers could adopt these principles in a more flexible way. In years of corn surpluses, for example, a greater number of fields could be put down to grass for grazing, and for longer periods. This would allow the farmer to build fertility for future years in what we might consider today a form of investment. It was a dynamic way of working the land based on smaller units: smaller fields, smaller farms and smaller settlements.

Most importantly, this system was resilient to change. While we might question its ability to generate the surpluses of the three-field system in a good year, in a bad year the greater diversity would allow it to resist the worst of the weather and disease. In the farming manuals of the eighteenth century this knowledge was known as 'convertible husbandry', and the environmental data from Exmoor was now allowing archaeologists to project its use back at least another one thousand years. This captured my imagination. If, I wondered, it had ridden the storms of deluge and drought for over 1,300 years, evaded the major landscape reorganisations of nucleation and the nineteenth-century Enclosure Acts, could the technique work for another 1,300 years? What really struck home, though, was the long-term adherence to a system that was resilient over and above a three-field system that, although riskier, had the capacity to provide greater profit margins.

ON A PRACTICAL level convertible husbandry intrigued me, since, if it was to prove successful, it must have involved a particular process. The devil is always in the detail and I found myself homing in on the one practical task that was required to make this system effective. Picture the scene: as a farmer, you've just harvested a crop of wheat from the field. As you're bringing the wagon loaded with sheaves into the farmyard for winter storage you notice that your yields are down on last year. The weather has been fair and other fields have produced well, so you decide that it must be a fertility issue and that it's time to give this field a break and turn it over to livestock for the time being.

Now, this isn't as simple as just setting a herd of cows loose in the field and letting them get on with it, for the cows would struggle to find any nutrition left in the stubble and bare earth of last season's wheat crop. What you have to do is plough the stubble up, work the soil down and sow in a grass crop – to give the cows something to eat while they're mucking their way to improving the economic capacity of your farm. Today, it's a matter of popping down to the local agricultural supplier's and picking up a bag of grass seed, plonking it in the sower and scattering it all over the field. It's remarkable how well even the most suburban lawn thickens up with a top dressing of grass seed acquired from the local garden centre. But where in the eighteenth century did you get the seed? More to the point, where in the eighth century did you get it? This seemed to me to be the missing link in an otherwise faultlessly explained system. The system of convertible husbandry projected back to the eighth century must raise the possibility that there existed a market for good quality grass seed, and therefore a series of agricultural processes constituted to produce it. If so, like wheat seed threshed from harvest stooks, it must have been threshed out of the grasses that made up the pasture.

There are numerous and complex permutations on how one might go about harvesting seed and hay and how it fitted into the farming year, but the main issue is that we must suppose that such a system would have been refined and perfected. The quality of the grass seed would determine the calibre of the sward, which in turn would dictate the health of the livestock and thus the wealth of the farm. So, in some ways, it was the critical part of the convertible husbandry system and I wanted to find out if there were any recorded methods for threshing grass seed in the Edwardian period, the eighteenth century and back as far as the eighth century. I'd only got so far with the history books, but I needed to get the verdict of a practical man.

It was for this reason that I'd broached the subject with Mr Mudge. He gave me a few examples of how a form of convertible husbandry had worked for him, and that it was in certain ways necessitated by the climate and relief of the west – particularly when working with some of the older and less disease-resistant strains of cereals, legumes and vegetables. It was when we got round to talking about the practical process of threshing the grass seeds from the hay crop that he leaned forward and gave me a canny I've-already-thought-of-that look. 'Ah well,' he said, 'when you come up to my place we'll have a look at a machine I picked up a couple of years ago. I think you'll find it very interesting.'

The Mudges' farm sat on the very edge of Dartmoor in a zone that for centuries had clearly been a battleground between the moor's natural propensity to re-wild and mankind's voracious appetite for domestication. To the east the tors of the inner moor loomed large on the horizon, dark clouds swarming around their summits. From there, vast expanses of open moor swept down towards the farm. As they drew closer, I could see the ruinous vestiges of past attempts at agricultural improvement: scattered walls and enclosures that grew increasingly numerous until they abutted the farm's inner fields. To the west the slopes dropped away to reveal the luscious valleys of the rivers Tavy and Tamar and the massifs of Cornwall in the far distance. This was a liminal place, a threshold between the bustling domesticity of fertile alluvium and the remote inhospitality of the thin acidic moorland soils.

The moment we arrived at the farmyard, Mr Mudge and I darted off to a small yard at the back of the farmhouse. It was crammed full of agricultural implements of all shapes and sizes and in various states of repair. I was in my element. We came to a box-shaped machine roughly the size of a van and covered in an old tarpaulin lashed down with baler twine. As he wrestled with the knotted twine he launched into a tall tale about how he'd acquired it. The suspense grew. Peeling back the tired canvas, he revealed what seemed to me an infinitely complex piece of machinery but one designed specifically for the job I'd been struggling to envisage. He gave me the how-it-works tour, and as none of the internal workings of the machine could be seen from the outside this consisted of him scampering around the contraption and pointing out the external wheels, belts, chains, levers and 'riddle' sizes that sieved and sorted the various seed types. In essence, it did the same job as a cereal threshing machine, only in sleeker form.

The hay was loaded into a hatch on top and dropped onto a conveyor belt, which then carried it towards a large drum. The tines on the drum thrashed it to separate out the seed from the stem. A second series of conveyor belts sped the thrashed hay away from the drum sump to an exit at the back of the box while a large paddle blew off the lighter material – the husk and chaff – allowing the cleaned seed to drop into the mechanically shaken sieves for sorting. For power, it could be belt-driven off a tractor or a steam engine and, if need be, there were adjustments to run it off a horse 'gin' – a horse-powered gearing.

This was exciting. Without taking my eyes off the machine, I stood back and pitched the sixty-four million dollar question. 'Why did you buy it? I mean, surely if you want grass seed you just need to go down to the agricultural merchant's and buy a bag of seed?' Mr Mudge just shrugged his shoulders. We both laughed at how it would certainly be more cost effective and time efficient to part with a few pounds sterling for a supply of first-rate seed from an international stockist, a variety perfected in laboratory conditions and designed to give the maximum output in terms of germination, growth and nutrition.

Looking back now I realise I never really got an answer out of him. Like me, it wasn't surplus and profit that interested Mr Mudge. It was resilience. In the first instance, he shared my desire to see equipment like this oiled up and put into action for pure interest and amusement's sake. But deeper down, this was a piece of kit that allowed him a degree of autonomy. If the global trade in seed stock dried up, or the agricultural supplier sold up and shipped out of town, he would still be able to keep the convertible husbandry system alive. This machine not only gave Francis Mudge the resilience and freedom to be his own man, forging a living on the edge of one of the most inhospitable places in the English landscape, it also enabled him to produce some of the finest roast beef I've ever tasted, from an agricultural system that was already 1,300 years old.

IN THE SHEEP farming models of old we have the perfect example of an economy we could learn from – a circular rather than a growth economy. Nothing grows eternally, after all, and it is arguably a philosophical flaw of our modern times that growth has become the ultimate objective for politicians and economists. A capitalist discovering the New World in the fifteenth century could perhaps be excused for feeling as if there was no end to the resources that might power the emerging European super states. We might even forgive the Victorians for believing that limitless natural resources confronted them as they delved deep into the heartlands of Africa and of the Indian subcontinent. Yet most economists and politicians persist in ignoring how this, the concept of perpetual growth, contradicts the accepted reality that the Earth's resources are finite. Hard as it is to accept, the days of industrial processes that 'take, make and dispose' may be drawing to a close and new economic models need to be explored. One such model is presented by traditional sheep farming, one of the purest of circular economies based on the most advanced, complex and renewable energies available to man: life, and the life-giving properties of sun and water.

The shepherd's year begins in October with 'flushing' the ewes. Flushing is the practice of setting the flock to graze in a particularly luscious pasture set aside for the purpose – often known as flushing meadows. Enjoying this rich and succulent fodder, as opposed to the thin dry grasses they will have lived off on the summer hills, the ewes then come into season in a good enough condition to improve their chances of conceiving twins or triplets. When they're ready, the ram is allowed to run among the flock to do his work – traditionally from 1 November – and they are together set either in designated pasture close to the farm or on a green fodder crop specifically sown for them to overwinter on. They might also be allowed the run of the stubble fields, where they can tidy up any lingering weeds and add some fertility back to the soil.

During the winter months their feed will consist chiefly of hay cut in the summer from the water meadows on the alluvial plain and fodder turnips left growing in the winter fields. Towards April is lambing time when, by keeping a keen eye on the flock, the shepherd should be able to intervene to avoid complications and keep the birthing success rate high. At this point the farmer is hoping that the first spring grass is coming through and the meadows that he 'haned' – locked up to leave ungrazed from Michaelmas time in late September – should, over any temperate periods of the winter, have developed enough length in the grass to provide early spring nourishment.

As spring properly takes hold, the ewes' milk will be at its best, the lambs will thrive, and by late May it will be shearing time at the farm. With the young lambs showing the first signs of independence, the weather turning to summer and other jobs requiring work around the farm, it's then time to send the sheep to the hills. Here, they will spend the summer grazing on the rich and varied sward of the downland, hillside and mountain. Even in areas where hill ranges are few and far between, vast stretches of marshland such as Romney Marsh in Kent and the Wash in East Anglia have long provided summer grazing for sheep and cattle. From the Pennines, Pentlands and Cheviots in the north to the South and North Downs in the south-east, from Dartmoor, Exmoor and the Quantocks in the south-west to the Black Mountains and Brecon Beacons of Wales, there is not a hill or mountain range, a marsh or moor in the British Isles that has not played host to this time-honoured practice of 'transhumance' – the seasonal moving of grazing beasts to summer pastures.

Transhumance is a pan-European practice that pre-dates our first historical sources. From the fjords of Scandinavia to the Alpine ranges of central Europe, from Spain in the west to the mountains of Transylvania in the east, transhumance – whether with sheep, cattle, goats or pigs – represents one of the oldest and most deeply ingrained agricultural practices of humankind. The norm was, and still is in certain areas, for whole sections of the community, usually the younger men and women, to spend their summers with their flock or herd, protecting them from predators and milking the cows, while the rest of the family stayed at the farm, tending and harvesting the crops. When the weather turned in the autumn and the nights chilled, it would be time to bring the beasts back to these valley farmsteads. This movement, a domestic migration from summer pasture to the security of the farm for winter, was an event with a deep and meaningful ritual significance. Popularly referred to as Samhain, today this seasonal feast is preserved in the annual ritual of Halloween or Bonfire Night.

In England, the basic premise of this ritual is that the animals driven down from the wilderness (a word which itself is derived from the Old English *wilder næs* – 'the lair of the beast') needed purifying, with any nasty spirits having to be purged from their bodies before they returned to the civilised surroundings of the farmstead. To achieve this, the flocks or herds were driven back to the safety of the farm between fires. Late autumn represents the time of year when all the summer foliage and plant growth has died back, and rather than leaving this material to decompose of its own accord, much of it would be collected up and burned, to avoid the spread of mould, rot and dampness. We all know the damage unraked leaves can do to lawns, smothering the grass over the winter months, suppressing its growth and in places killing it off entirely. The same holds true for a farmer's precious pastures and swards. There was also a more general need at this time of year to prepare a way to the boundaries and to begin thinking about the winter work of laying hedges, digging ditches and drystone walling. In this time of general clearance, burning all the detritus provided a useful source of potash for the fields around the farm, as well as ritually driving the evil spirits from the bodies of the returning livestock.

And so the cycle would begin again. This was a cycle that converted infinite renewable energy – sun and rain – into meat, wool and soil fertility (and of course, bone, horn, hide and all the other parts of the animal that wouldn't have been wasted). It works, and has done so for millennia. The key to its success is that during the good times – the summer, when the weather is warm and everything is plentiful – the sheep are placed in the most testing environment of the region, up on the hills or out on the marsh. There they have to work hardest for food, foraging on coarser pastures. The vital asset in this cyclic economy therefore becomes lean and hard-working rather than fat and lazy, which they would have become had they been allowed the luxury of highly nutritious lowland grasses during the warmest months of the year. So when the time comes for them to be brought down from the hills, the sheep are in good shape to confront the hardships of winter – the bad times – and to make the most out of the fodder they're given to deal with the wet and cold. This, then, is the rationale behind the circular economy: make your assets lean in the good times, in order that they are better equipped to weather the bad times. This is a system that not only balances its inputs against its outputs but is designed so that the cycle is self-sustaining. Equally, it is not carried out in opposition to or in disregard of other processes that must take place on the farm. For, in the case of sheep farming, removing the beasts to the hills prepares them for winter and allows crops to be grown on the farmstead.

Modern sheep farming, and livestock farming more generally, has largely dispensed with these principles. Instead of integrating their livestock into a wider system, farms have become little more than processing plants where cheap imported animal feed is converted into meat. This is a system of capital investment in which feed becomes profit in the form of meat. The more you pump into your livestock the quicker they fatten, and bigger animals on a faster turnaround – a particularly popular strategy with cattle – can make you money faster at market. When the market is with you, you stand to boom. But with the market against you, and limits to how much your business can exponentially grow, bankruptcy is never far away.

The extreme and logical conclusion of this system is one where more beasts are kept in sheds in ever smaller spaces, pumped full of fat-rich food and growing abnormally quickly so that a return can be gained as quickly as possible. The real point here is that this model is entirely dependent on processed feedstuffs derived from source crops such as cereals, roots and brassicas that could otherwise be fed to humans. It also requires a level of capital investment predicated on a supply of cheap fossil fuel. The shepherd in the transhumance system – the circular economy – produces at a sustainable and more resilient level. The yields may not be as high, but the system requires the absolute minimum of energy inputs in the form of sun, rain, and the inherent ability of nature to reproduce. This is the *woruldcræft* of Ælfric's *Colloquy*, the integration of arable and animal husbandry, the balance in any landscape between seed and sward, in order that one can support the other and vice versa. We abandon this most fundamental of crafts at our peril.

11

THE *OXNA MERE*

ONE OF MY favourite places to walk is the chalk downlands rising up to the north of the Vale of Pewsey in Wiltshire. Like most people who enjoy rambling, I see walking as a means to clear my head. I call it 'walking out'. In the same way I would iron out the creases in my laundered shirts on a Sunday evening, I would walk out the creases of my cluttered mind on a Saturday afternoon. Every now and again I like to take on new landscapes, and spend most of my walking time navigating and taking in the sites – learning the lie of the land. But to properly walk out you need a landscape that your feet and legs are familiar with. You can let them take over, allowing your brain to disengage and concentrate on the defragmenting of your internal hard drive, sorting, ordering and creating space for processing the week ahead.

Back in the early 2000s I used to visit this part of Wessex with almost monthly regularity, choosing a different circular route with every trip. It was a turbulent period of my life and a time when I needed a landscape I could trust. As part of a mid-twenties identity crisis, I was very purposely developing a relationship with the chalk downlands of southern England. From where I grew up in Sussex, on the borders of Pevensey Marsh, the gentle curves of the South Downs could be made out in the far distance to the west, a constant presence during my childhood as I played in the fields, spinneys and meadows that bordered the marsh. As soon as I was old enough I would bike my way over to them, following the medieval lanes that snaked across the flatlands between hillocks and spurs of high ground, then clamber up the steep scarp slope. I would spend long days exploring the distinctive ridgeways, deep combes and sweeping summits on what is probably one of the most renowned stretches of chalk downland in southern England.

One would think the term 'downland' implied an area of lowland, in opposition to 'upland' or higher ground. But in one of those curious tricks that the English language can often play on you, the 'down' element is derived from the Old English *dune*, one of the Anglo-Saxons' many words for describing a hill. These are, therefore, hill lands, and although the soft chalk geology responsible for their curvaceous beauty can be found making its way north through eastern England, producing substantial outcrops in Norfolk, Lincolnshire and Yorkshire, it is most abundant in the southern counties of England: Kent, Sussex, Hampshire, Surrey, Wiltshire, Dorset and Berkshire. The downlands have thus come to embody a sense of rural southern England immortalised in the prose of Edward Thomas's *The South Country*, W. H. Hudson's *Nature in Downland* and, perhaps most famously, Richard Jefferies' *Wild Life in a Southern County*. When I moved to London, I very quickly tired of the stifling noise and bustle of the overcrowded city and would often sate my desire for the downland hills of my youth via a weekend trip to the chalk. With tent, blanket and map packed, I'd put on my walking boots and head for the country – embarking on a relationship with a landscape I continue to enjoy to this day.

Over time, I came to favour the chalk downlands of north Wiltshire, and a direct line from London Paddington would take me to the heart of some of the best downland country in southern England within a matter of hours. For me, the Vale of Pewsey, a kind of no-man's-land between the better-known Stonehenge landscape to the south and the Avebury landscape to the north, became a particular favourite. It was quiet and gentle in the Vale, exposed and glorious on the downs that bounded it to the north and south. It was a sensational landscape for an archaeologist. Whichever route you walked would take in an inordinate number of archaeological landmarks, providing vistas to countless others.

The most enigmatic monument of the Pewsey downs was a vast linear bank and ditch that stretched for a distance of around twenty miles across the top of the Vale. Known as Wansdyke, a modern contraction of the Old English *Wodnes dic* ('Woden's Dyke'), this was a majestic edifice with a ditch that in places reached depths of over ten feet and a bank that swelled to heights of over thirteen. I would often seek out Wansdyke in order to incorporate a substantial stretch of it in my walk. On the warm sunny days of summer I would traipse along the crest of the bank, providing a vantage point from which to enjoy the surrounding landscape while allowing its sinuous form to guide my feet. In the harsher wet winds of the winter months I would tuck myself into the ditch, using the bank to shield me from the prevailing south-westerly squalls.

No one knows exactly who built Wansdyke and for what purpose. In facing north, it was undoubtedly intended to protect the land and people to the south, and the period of antagonism between the aggressive hegemony of the Mercian kingdom and the emerging strength of the fledging kingdom of Wessex seems a likely context. But until any further archaeological investigation is undertaken, it shall remain one of Britain's largest and most obscure monuments. Woden crops up in a number of other locations on the downs to the north of the Vale of Pewsey where the Old British Way, the ancient name for the present-day Wessex Ridgeway, rides up from the Vale and passes through Wansdyke. The exact intersection was known as Woden's Gate, the head of the valley the route continued along was once Woden's Dene, and the long barrow that marked the summit of the downs immediately overlooking the Vale was originally known as Woden's Barrow. I knew all these names – and the association of these places with Woden – because I'd read the ancient land charters for the area, drawn up in the age of the Anglo-Saxons.

I envisaged the few ramblers and dog walkers I encountered on these windswept hills to be largely oblivious of the lost knowledge relating to this cult landscape and its dedication to the all-powerful Saxon pagan god. And it would seem that as early as the sixteenth century Woden's Dyke had become abbreviated in such a way as to obscure the connection to Woden. The antiquarian John Leland remarked on how the local people believed it was named so because the Devil had built it on a Wednesday – 'Woden's day'. For Woden's Barrow it was less the case that the connection slipped casually from folk memory over time and more an example of a fervent Christian community seeking to update the significance of the place, converting it to their own religion in the present rendering of Adam's Grave.

During my early visits I baulked at this unashamed Christianising imposition on such an ancient landscape. But with every walk I mellowed and almost warmed to the evolution of names, languages and beliefs and the necessary loss it entails. There is some logic to the choice of Adam and not some other biblical protagonist or saint. Observing the shape and form of the Vale of Pewsey as it appears on a map, and the way it funnels into the Avon, I've always thought that it takes the form of a uterus – or at least the diagrammatic uterus of my schooldays human biology textbooks. It seems fitting for Adam, the first man from whom all human life has its genesis, to sit so prominently overlooking the Vale and the life-giving springs that go on to feed the Wiltshire Avon. This is Wessex's heartland river. It flows south to the sea via Woodhenge, Stonehenge, Amesbury, Old Sarum and Salisbury. There is, therefore, a sense of *beginning* in this landscape. The name may have changed but the sense meaning has remained intact. For the later-medieval Christians it was Adam, the first man, who should most fittingly have been interred in this barrow. But for the pagan Saxons it was their progenitor, Woden, who most appropriately commandeered this ancient burial site.

One landmark I seemed always to take in on my rambles through this landscape was the *Oxna Mere*. Far less glamorously mythological and much more agricultural in its nature, the 'Oxen's Pond' was nothing more than a large saucer-shaped depression tucked into a dip in the downland at the head of one of the many combes that perforate the scarp slope of the Vale. The distinct change in the flora in waves of radiating halos around the saucer suggested that this place was no stranger to the occasional phase of drought. Over centuries, a resiliently plucky vegetal community had carved out a life, the almost luminous yellow of young pond sedge grasses in stark contrast to the surrounding deep green of the downland sward. Only in recent years have I come to understand why I gravitated to this relatively innocuous – and distinctly profane – monument of my dear chalk downland. I adore the myths, the legends, the belief systems of the ancient people and the burial mounds and fortifications associated with their gods and leaders. But I guess my real interest lies in the resolutely pragmatic function of this pond, providing water for livestock on what was otherwise a parched hillscape. But most enigmatically, why construct a pond so far away from any viable supply of running water?

I knew that for as far as records stretched back the downs of southern England had been used for the grazing of livestock in a form of transhumance. Up here on the exposed hills there were enormous acreages of grazing to be had for sheep and cattle in the summer months. This was the perfect time to exploit the rich feeding grounds on the hills, while at the same time keeping livestock away from the crops growing in the valley. However, the chalk downlands are famous for their dry valleys and for being incredibly arid at the height of summer. And livestock, especially cattle, need to drink regularly to maintain a good healthy condition. So I could see the logic of these ponds: construct a watering hole up on the downs and it could save the herdsman the onerous daily task of driving his herd down towards a valley water supply. This took them dangerously close to luscious meadow grasses (preserved for winter fodder) and cereal crops (for human consumption), but it saved an arduous journey that would cause his livestock to expend energy that could otherwise be stored up for the winter. The problem is, in all my countless rambles across the chalk downlands of southern England, I had only on a handful of occasions found one of these ponds to contain any water. But there were so many of them that they must, at some point, have worked. It's one thing building a pond out of entirely natural materials, but so far from a spring, or running water, it seemed to me quite another thing to find a way of filling it and then maintaining a sufficient water level. This truly was a lost knowledge.

I knew this one was called the Oxen's Pond because it appeared, alongside all the Woden-related landmarks, in the ancient Anglo-Saxon charters that recorded the exchange of property in the Vale of Pewsey and the downlands beyond. The charters for this particular area dated to the tenth century, but the topography they describe is very likely to have already been hundreds of years old. As I sat by the pond taking a well-earned break from my walk, I would imagine the Saxon herdsman frequenting this watering hole at the height of summer, busily goading and shuffling his charges in order that every animal could take time to quench its thirst. It wasn't until I discovered that right up to the twentieth century people were still creating these ponds that my interest was really awakened. Could it be that an unbroken tradition of excavating and maintaining ponds on the downs was in operation for a period of over a thousand years? Better still, if we could study the ethnography of late-nineteenth- and early-twentieth-century pond construction – which is comparatively well documented – could we back project some of the practices, methods and techniques onto the pond makers of Anglo-Saxon England and better inform ourselves about the Old English ways of managing the chalk downland landscape?

ONE OF THE reasons we know quite a lot about pond making on the downs in the nineteenth century is because of the interest in them sparked by Arthur John Hubbard and George Hubbard's *Neolithic Dew Ponds and Cattle Ways*, published in 1905. A spate of correspondence, papers and reviews in the pages of the *Journal of the Royal Society of Arts*, *Nature* and the *Geographical Journal* from 1907 to 1911 led to the publication of other books on the subject – in particular, Edward Martin's *Dew Ponds: History, Observation and Experiment*, published in 1915. Through much of the writing of this period there is a trace of antiquarian interest in these ponds with some speculation that they related to very ancient times and were contemporaries of the ring ditches and barrows that peppered the downlands in the prehistoric period. But the main sticking point for most contributors was not the great antiquity but rather the designation 'dew ponds', a term first used in the mid-nineteenth century but quickly passed into popular parlance among the archaeologists and topographers of these hills in the twentieth century. The implication in this name is that these ponds were filled and replenished through the capture of dew – microscopic droplets of water that form on chilled surfaces when atmospheric vapour condenses.

It's an attractive concept that, by some mysterious alchemy, these dew ponds replenished themselves overnight. And at first glance it seems the most logical explanation for how purposely built ponds so far away from running water could have proved useful to stock-keepers. To go to the effort of constructing them in the first place, there must have been some means by which they were filled with water and remained so for significant periods throughout the year. As anyone who has camped out on the downs in the height of summer will tell you: wake up first thing in the morning, start tramping around outside your tent and very quickly your feet will be saturated by the copious dew that forms on the blades of the tightly grazed downland grasses. Naturally, if the water in the ponds was at any point colder than the surrounding land and air, a disproportionate amount of dew would end up in the pond, helping to diminish the effects of daytime evaporation. This, at least, was the view of the Hubbard brothers, who attributed the construction of the ponds to the Neolithic period, linking them with the cattle economy of ancient prehistory and advancing the theory of the non-conducting properties of straw playing a significant role in the insulation of the water from the rising temperature of the ground below.

Over time, the lack of scientific rigour in the work of the Hubbard brothers has caused many of their assumptions to be found wanting. Their most severe critic, Edward Martin, argued that the primary means whereby these ponds were stocked and replenished was through rainfall, and not through the capturing of aqueous vapours. The Hubbards had been too keen to play up the role of condensation in part because of the attractive lure of magical self-filling ponds but also because of a well-meant desire to export this most ancient of technologies to other parts of the world where water – and rainfall – was in short supply. Yet in the overcooking of their own case they had perhaps courted a slight overreaction in the meteorologically savvy readerships of the *Journal of the Royal Society of Arts*, *Nature* and the *Geographical Journal*. The problem is, of course, that if these ponds relied so heavily on rainwater, what use would they have been in the height of summer, the season of least precipitation, at precisely the time when such drinking facilities would have been most in demand from the flocks and herds that roamed the downland?

In so successfully debunking the work of the Hubbard brothers, Martin and company may have been guilty of throwing the proverbial baby out with the pond water, for it remains the case that if such ponds had no other means of sustaining water levels during periods of hot and dry weather, they would have evaporated at the very time of year when the farming system of southern England would have had most need for them. The Hubbards' detractors essentially leave unanswered the curious business as to why people have been building ponds for over a thousand years on the downs in positions so far away from a consistent water source.

It's clear that these constructions operated on a much more sophisticated understanding of meteorology than we give our Anglo-Saxon, and perhaps Neolithic, forebears credit for. There's no doubt that rainwater played a large part in the filling of these ponds, and the first facet of their intelligence should surely be their location within the landscape, at depressions sitting at the heads of dry valleys, accentuating the natural lie of the lands to make maximum use of a rainfall catchment area. But this is not always so, and certainly not the case with the earliest documented ponds. The Oxen's Pond, for example, would be much better placed a short distance to the north-east, some sixty-five feet downslope in a natural hollow that would have collected water from a much wider catchment area. That is, if rainfall was the main consideration. Instead, it's perched almost at the very summit of Milk Hill, at a height of around 950 feet above sea level. What's more, our south-westerly winds are those that most commonly bring in the warm and wet weather from the Atlantic Gulf Stream. Yet the *Oxna Mere* sits on the leeward side of the hill (as opposed to the wetter windward side). If anything, it's placed on the driest part of the hill. The same can be said of the two ponds that sit on Cannings Hill one and a quarter miles west-north-west at the same altitude. And at another of my favourite haunts, Hackpen Hill, the highest point on the Marlborough Downs, there is a pond sited in a similar location – on the leeward side of the summit at a height of around 885 feet above sea level.

Altitude must be a factor in the positioning of these ponds, and the inescapable conclusion, if they are designed as functioning waterholes, is that condensation plays a greater part in their replenishment than does rainfall. Differences in the construction techniques of ponds in different areas may go some way to help explain this phenomenon. To begin with, there may be an element of truth in the Hubbards' assertion that straw is used as a non-conductive layer in order to shield the pond water from the rising temperature of the ground long enough for it to sustain a dew point. Where we have records of pond making it is in Wiltshire, places like the Vale of Pewsey, the Marlborough Downs and Salisbury Plain, where straw is advised. In other areas, like Sussex and south Dorset, straw doesn't seem to be as much of a requirement in the construction process. The reasons for this may have something to do with the proximity of these ponds to the coast and, in particular, to the sea mists that blow in at all times of year. It's important to register the capacity of salt to retain water, and to think of these sea mists bringing in particles of salt, gorged with vapour, as they gently roll across the upper reaches of the downs.

For a period of about two years I used to commute along the A27, a road with undoubted medieval origins, which passes through the towns and villages strung out along the slopes of the South Downs. As a jobbing archaeologist, I used this arterial trunk road to get from one end of Sussex to the other, and would often pass large stretches of chalk downland on the seaward side of the road. During the summer months I frequently drove through dense mists brought in off the early morning swell. Even on the hottest days of the year these mists would cling to the peaks of the chalk hills. The porous chalk, chilled by the clear skies of a summer night, would ensnare the moisture that had risen from the warm night sea and drifted overland on the breeze. It may be because of this proximity to moisture-laden sea air that the ponds of the South Downs are traditionally recorded as not including straw layers in their lining – for there was no need to provide the extra insulation that was deemed important in the more inland areas of Wiltshire.

At the turn of the nineteenth century the Marquess of Aylesbury, concerning his lands in the Vale of Pewsey, insisted in his leases that in the maintenance and construction of ponds, three successive layers of straw and clay were required to build up a composite pond lining of two feet and six inches thickness. This stipulation, with its institutional backing, could represent an innovation characteristic of the agricultural improvements in the eighteenth century. But it might just as well have its origins in the ancient and learned practice of the estates in this particular area. In each case the straw layers, comprising carefully laid 'best wheatstraw', needed to compact down to three inches deep, with each clay layer being puddled and then compacted to a depth of eight to twelve inches.

The introduction of organic material might be questioned on the basis that, in rotting down, it would create a void that would impact on the structural integrity of the pond. However, the pond makers of old must have been familiar with the science of anaerobic conditions, and of how in environments starved of oxygen organic matter can be preserved indefinitely. What is interesting in this example is that in other areas where straw is used it could feasibly be interpreted as merely the dividing membrane between a smooth clay compaction and rough rubble courses – to stop the former being penetrated by the latter. But in the Marquess of Aylesbury's requirements it was used as a functioning entity between layers of exactly the same material, which therefore suggests a role as an insulator.

The form of the ponds also seems to be important. They are invariably shallow dish shapes rather than deep straight-sided cisterns, which is strange because if rainwater catchment was a primary concern surely a greater depth would secure a greater level of storage. It may be that deep pits of this type presented too much of a drowning hazard to thirsty animals – although quadrupeds tend to be cautious in the placement of their feet when unsure of the surface beneath them. It could also be that shallowness allowed for more volatile temperature change within the water and therefore more frequent opportunities to achieve a dew point.

It could also be that waterproofing a deep cistern pit was problematic. Chalk bedrock is famously porous, and having lived on it for much of my life I never cease to marvel at the ability of the overlying shallow soils to dry up in a mere twenty-four-hour period after even the most sustained deluge. Obviously some kind of water-resistant membrane is required. Today, this is achieved through the purchase of a purpose-made pond liner – available at all good home-improvement retailers. But in an age before plastics, polyurethane and concrete, a very particular craft is required to create waterproof layers from the bare earth. The technique is known as 'puddling' and involves the repeated stamping on very fine-grained materials such as clay under wet conditions. This is something that can be done with one's own feet, and in constructing a rammed clay floor some years ago I spent the best part of the job marching on the spot in an oversized bucket of clay, working it down until it was brilliantly smooth and pliable. It was heavy going and exhausting, and at the end of a long week I had thighs of iron. Of course, it's always far better to get someone else to do the hard work for you, and using the weight and natural plodding action of cows is a well-known means to meet your clay-puddling needs.

The practice reaches its high point in the construction of the canal network of early industrial Britain when large herds of cattle would be run through the empty canals to puddle and compact the clay lining to improve water retention throughout the network. But cattle in herds are not easy to control, and to achieve this rather specific task with any efficiency, it would be to oxen – beasts of burden – that one would have to turn. I'm very much of the mind that the *Oxna Mere* takes its name not from the livestock using the pond as a watering hole but from the draught beasts that were used to construct and maintain its waterproof lining. And I would dearly love to excavate this pond to see what exactly it is made of. At 950 feet above sea level, on the very top of the Upper Chalk, as the geological stratum is called, it's some distance from the nearest source of clay. It may well be that the Oxen's Pond was constructed using puddled chalk – a method recorded in nineteenth-century Sussex. This was new to me. While familiar with the plasticity of clay, I hadn't previously considered that chalk could be puddled to create a waterproof liner. It can, of course, be ground into a very fine powder, and it stands to reason that such a powder, recompacted under heavy feet, might re-form into a solid layer that could prevent the passing of water.

There may have been a difference in the way a pond was constructed depending on the animals it was intended to serve. The preferred method for a cow to quench its thirst is to tentatively wade in to shin height and drink deeply. Sheep, however, tend to tiptoe to the water's edge and lap away. The concern with the heavier of the two is that repeated accessing of a particular location in the pond will result in rutting and the eventual penetration of the clay liner. Consequently, some methods of construction involved lining the top of the clay with either chalk or flint rubble to create a cobbled surface to protect the soft underlying clay. Other materials have also been recorded in the construction of these ponds, and a useful means of preventing the invasion of the waterproofing clay layers by burrowing worms is a layer of quicklime – a caustic substance that aggravates their mucus membrane and dissuades them from burrowing any further.

Concrete started to be used in the mid-nineteenth century as industrial processes of extraction and transportation made it much more widely available in rural areas. And what's interesting about the experiments carried out by Edward Martin is that, as a quick-fix solution to creating a waterproof pond, he used asphalt, tar and pitch to replicate the historical ponds whose functioning he was so keen to elucidate. Black, dense, heat-storing and thus behaving in an entirely different way to the natural materials available to the traditional pond maker, it's no wonder he drew the conclusions he did.

The implications of using puddled chalk were important to me in the context of the *Oxna Mere*. Ultimately, its significance lay in the simple revelation that if you had the knowledge and skill to puddle chalk, you could create a watering hole using materials sourced entirely from the hilltop. In turn, this facility would make an important contribution to the methods of husbandry used by the valley community in that it enabled them to exploit valuable resources of summer grazing in a more effective manner. This is the kind of thing I get excited about: resourcefulness on a level almost inconceivable to the post-industrial pond maker whose favoured materials were concrete and asphalt.

THE ISSUE AS to whether these ponds were Neolithic in their construction is another matter altogether, distinct from whether they were intended as dew ponds or rain ponds. There is a scenario whereby we might be able to prove their ancient Stone Age origins. It would require excavating a pond, finding an antler pick associated with some kind of lining layer and radiocarbon dating the antler to the period 4,000–2,500 BC. An unlikely but not unimaginable scenario. Like roads, ponds are one of those landscape entities that are notoriously difficult to date because they're in constant use. The best place to put a route between A and B four thousand years ago was invariably the best place to do so in the very recent past. The best place to put a pond in the Neolithic was invariably the best place to have a pond in the Anglo-Saxon period. In both cases, why go to the effort of creating a newer version when with far less work you could renovate, modify and improve the existing one? Of course, in improving an existing pond, you likely reorder and disturb archaeological deposits from an earlier period. It stands to reason that John Gould, recorded as the parish pond maker for Ebbesbourne Wake in the years 1851 and 1907, must surely have returned to the same ponds to maintain and rework their linings, as his forebears will have done before him. You'll see it written that there is no evidence to suggest that dew ponds date back into prehistory, but be wary of such bold assertions. There is no evidence to suggest that they don't.

The *Oxna Mere* proves that some of our ponds can at least be dated back to the late Anglo-Saxon period, but whether they were new constructions in the landscape at this time or reworkings of more ancient ponds is difficult to say. A couple of years ago I resolved to undertake a study of dew ponds in a particular area of Wessex in a bid to try to understand how they functioned in relationship to the wider early medieval landscape. The area I selected was on the border between Wiltshire, Hampshire and

Berkshire and consisted of a large expanse of downland that, but for the headwaters of the Bourne Rivulet, was essentially devoid of any water sources. There are numerous ponds referred to in the Anglo-Saxon charters for this area, and the way they're described says something about how they were being used. A lily mere, for example, suggests a pond that holds water long enough throughout the year to support a colony of lily plants. A dry mere suggests one that has long since fallen out of use, and perhaps indicates a pond of greater antiquity.

The pond name most indicative of a working grazing landscape is the one that gives its name to the wider estate that is the subject of the grant recorded in the charter. It's called buttermere, and there can be no clearer indication that this pond, along with the other active ponds in this area, was supporting a form of dairy farming. Plotting the references to ponds in the ancient charters for this stretch of downland – as well as later recorded ponds that include the 'mere' place-name element, such as Ashmore Pond, Limmer Pond, Rushmore Pond, Wadsmere Pond, Wigmoreash Pond – reveals an even distribution across the study area, all occupying the higher ground summits and ridges of the Upper Chalk. The impression given by this distribution is one of a meticulously planned landscape that supported a number of herds and flocks. The estimations in the nineteenth century of how long a pond could last before it needed reworking were between fifty and seventy years. The suggestion, therefore, is that they were being reworked, relined and maintained in the Anglo-Saxon period. But that some are recorded as dry in this time might suggest that the Anglo-Saxon exploitation of this hill hadn't reached the extent of an earlier age. Justifiably, we might project the origins of these ponds back into the Romano-British period, and that many of them were being used as landmarks to set out the boundaries of Anglo-Saxon land units supports this assertion.

ONE POND THAT was almost certainly not of Romano-British origin and unlikely to have been of Anglo-Saxon date was New Pond, which I had the dubious honour of living alongside for a period of ten years. It lent its name to New Pond Field, a place first recorded in an estate map dated to 1650. This useful etymological clue to the pond's origins told me that it was there in the year 1650, but exactly how 'new' it was at this date is anyone's guess. Our cottage had been built in the early twentieth century in a triangle of redundant ground between an ancient hedge boundary, a farm track and the pond. We were located in the centre of what had been a large medieval hunting forest, and I rather suspect that the pond's construction was related to a change in the function of this landscape in the decades running up to the date of the map. As sheep farming took off in the sixteenth century and demand for timber saw traditional forest cover reduced, the construction of a pond at this location would have provided a useful watering hole for the introduction of livestock – particularly sheep, but perhaps even cattle.

During the course of my pond-side residency I had the pleasure of watching a historical pond's slow deterioration over time. The ten years I spent observing an almost imperceptible change to this haven of wildlife represented a mere chapter in the pond's longer-term chronicle, but it was an important one. During this period it reduced from a receptacle just about capable of holding water to one that drained itself away during the drier months of the year. It was a period of accelerated decay. Once it started to dry it was weakened. Cracks emerged, weed seeds penetrated, and the clay lining was riddled with the taproots of dock, burdock and hogweed. I would have loved to step in and reverse its inexorable journey to desiccation, but I had neither the time nor the money. I settled for meditating on my helplessness, while simultaneously revelling in the volatility of the micro-environment the pond's construction had created.

This pond was not a dew pond. It was designed to trap rainwater from the valley and had been located in a place well suited to do this, while at the same time situated equidistantly between the main farms of the estate. During the wettest times of the year it would fill at considerable speeds since the roads snaking down from the chalk ridges towards the pond – and my cottage – ran like rivers. A steady fall of rain at daily intervals kept it topped up, but in periods of sustained heavy rain it would burst its banks, encroach on our garden to the immediate north before flowing off to the south, onto the road and the fields beyond. It was this influx of rain that created such a volatile environment, both in the waterlogging of the pond and its immediate surroundings and in the carrying of sediments from the surrounding ploughed fields.

In some ways, another change in function for this landscape in the post-war period meant that this pond's days were numbered as a useful stock-watering hole. Much of the downland in southern England was given over to the growing of cereal crops, a scheme created and subsidised during the Second World War and one that has continued – effectively in a subsidised state – ever since. It may come as a shock to many that less that 5 per cent of indigenous grasslands from the 1940s have survived this onslaught. But the net result for New Pond was that it was no longer needed as an occasional watering hole. The rains that had been its provider for over two hundred years now served only to bring in tides of fine sediments derived from the destabilised ploughed soil of the surrounding fields, laden with a potent mix of artificial fertilisers and chemical insecticides.

The flora of New Pond was thus of the most aggressive kind. When I first arrived at the cottage a dense curtain of nettles surrounded the pond, embedded into the broken and (artificially) nitrogenous soil of its bank. One hot July evening, armed with a fallen branch, I slashed my way through nettle plants, in places over six feet high, and burst through onto a pond which, thanks to a number of overhanging willows and a screen of ash and cherry trees, was just about holding a shallow saucer of dark water. But it was struggling, and it was obvious that the pond base, where exposed, had hosted a range of plants. The following year an invasion of dock seeds brought in on pluvial deposits germinated in the pond at the beginning of a warm autumn, after a long summer when the pond had run quite dry. The young plants rode the mild winter, and after a dryish spring their rapacious growth rate had an irreversible effect on whatever water-retention properties the clay lining might have had.

I considered that this was the pond's Waterloo moment. There was no going back after this onslaught. As an aside, when I moved to this cottage there was a dock plant growing between two concrete slabs on the path just outside the back door. Determined never to use chemical weedkillers in my garden, and incapable of lifting the slabs to extract the root by hand, I resigned myself to hoeing off the top of the plant three or four times a year to keep the back of the house in neat shape and to avoid it maturing, pollinating and spreading its seeds into other cracks on the path. 'One year's seed, sixty years' weed', I remember an old farmhand saying of the dock leaf plant. I vividly recall the day I left the cottage and glanced down at that dock. In victoriously withstanding the hoe blade for so long, it had essentially outlived my tenancy – the bastard. I suspect it's still there today.

I felt the same pang of resignation on the morning I realised that the young plants germinating on the pond floor were dock leaf. I'd never get rid of them, I thought, and would just have to learn to live with them and their copious production of seeds each year invading my vegetable patches and flower beds. However, in a quirk of fate, and a poetic illustration of how even the most resilient species can suffer irreversible setbacks, one November the pond filled and held water until late April. By mid-May it was dry, and I realised that this sustained period of waterlogging had effectively sterilised the soil – saturating to death any existing plants and residual seeds. Squatting in the base of the dry pond, rubbing the rich humic soil between my fingers, I saw my opportunity. If I could sow seeds of my own choice, I could potentially steal a march on any would-be early colonisers. Damn you, dock leaf. It was payback time.

And so I raced to the local agricultural supplier's and picked up a twenty-five kilogram bag of sunflower seeds. I broadcast sowed them liberally across the dry pond base, raked over a thin tilth and, with my antique lawn roller, pressed them down to get an intimate contact between seed and soil. The rains came, warm and softly, and the temperature picked up. Everything went as planned. One or two wood pigeons cottoned on to what I was doing, but with patience, and the judicial use of the airgun, their unwelcome attentions were mitigated. As the nursery leaves of young sunflower plants emerged in early June, I celebrated with some home-made elderflower beer – and pigeon pie. By mid-October an awesome sight presented itself. My pond teemed with a crowd of sunflower plants proudly turning their heads in a blaze of yellow-golden pinkish-reds as the sun arced its way over the late autumn sky. That year my bees – as well as a host of other native insects – gorged themselves on this extra source of pollen. During the winter months, chaffinches, greenfinches and – a personal favourite – bullfinches became regular visitors to the pond to stab their beaks energetically at the flower heads. For every kernel they cracked and every seed they swallowed, a further two or three would scatter into the thick undergrowth below. Next year's crop was, I believed, ensured.

Of course, you can imagine what happened. The following spring an impressive number of sunflower seeds germinated in the pond, but a deluge in May waterlogged the delicate immature plants and killed all but one of them. This lone plant reached maturity and seeded itself, the proud survivor of a literal golden age. Cowed by my experience I stepped back from the pond, again retreating to the role of passive observer as the deluges destabilised, silts amounted and species came and went. In the construction of this pond in the first place, a biodynamic environment had been created – and one alien to the chalk downland. But now prone to running dry, its volatility was amplified. It was like a festering open sore in the landscape, a constant no-man's-land in the battle between the warring propensities of the most vigorous earlier colonisers.

It wasn't just the dramatic change in the flora that highlighted the plight of this pond. It was the change in bird life over the course of my first few years at the cottage that first suggested to me that the pond's condition was not static and that it was in a process of degradation – and that this decline in its ability to hold water all year round was a fairly recent phenomenon. When I first clapped eyes on Roger, a moorhen which I could tell from his markings was male and mature, I wondered how this flightless bird had come to live so far from a reliable source of water. We lived a good mile and a half from the nearest stream, and it was only after repeated sightings of this shy and retiring bird darting amid the undergrowth around the pond that we realised he was a resident of this part of the valley. The gamekeeper's wife told us he was also occasionally sighted up by the farm buildings, where vast corrugated tin and asbestos roofs created enough run-off to feed a number of ad hoc ponds around the complex. It was at this point that he was given a name – after my favourite James Bond

– and as we were relatively new to the area, and devoid of other neighbourly goings-on to gossip about, his movements became a frequent subject of conversation in our household.

'I saw Roger this morning,' my wife would say of a Sunday morning.

'Oh, really, darling,' I would reply, not lifting my gaze from the sports pages of the Sunday papers. 'Where?' I'd ask casually, interested nonetheless.

'He was foraging at the base of the crack-willow, but headed off towards the hedge bank.'

I'd look up, raise my eyebrows in acknowledgement of Roger's business and then return to the papers.

Roger died three and a half years after we'd moved to the cottage. I found him on a cold January morning. Bless him, he'd perched on the compost heap to make the most of the residual warmth from the decomposing mulch. I picked him up and brought him into the sun where, in the wintery light, his bright orange bill and the vibrant green of his legs were brought into magnificent contrast against the glossy black of his plumage. I buried him at the back of the garden, overlooking the pond that had been his home for so many years. Like the lone sunflower, Roger was a survivor, the last in a line of moorhens that had once made this place their home. But this habitat was changing, and a bird that needs water all year round, that forages on aquatic species, could no longer call it home.

Another result of the slow degradation of the pond's fabric was the change in breed of the family *Hirundinidae* that occupied the eaves of the cottage on the south-west-facing wall. When we arrived, a colony of swallows was in residence, raising their offspring in a series of mud-cup nests clinging to the walls immediately under the roofline. From here, they would fly out and catch insects on the wing, drawing vast figures of eight in the skies above the cottage, screeching in delight as they ducked and dipped in the twilight sun. Masters of the skies. But by the time we left the cottage, house martins, the swallow's smaller and slightly less glamorous cousin, had commandeered these nests. Characterised by their giggling chatter, the house martins seemed to me to be a more amusing flycatcher, but I realised after a number of years why they were beginning to dominate over the swallows.

There seemed to be, at least in my immediate locality, a fundamental difference between how each breed collected mud and water to build their nests. In the early years, I used to love watching swallows swooping down on the pond and scooping up beakfuls of water on the wing to mix with the dirt already stored in their gullets. As the pond began to dry up for longer periods in the spring, the opportunities for wing-borne water catching were fewer and the swallows would fly long distances to get the necessary water to make their nests. The house martins, on the other hand, seemed to be more adaptable to the situation and would make use of the slightest puddle, landing on its fringe and pecking up the water-laden silt. Critically, they were prepared to ground themselves to do this.

Now, I probably gave the house martins a bit of a leg up in this situation, allowing them to make maximum use of their willingness to work with the most minimal of mud and water supplies. There was a particular puddle that occurred in the road at the point where our car turned into our drive – a location that was in direct line of sight of the house martins' nests. At times of the year when the main pond was so dry that even a moderate shower failed to saturate its base to the point of providing sitting water, this tarmac-lined pond would trap enough water for the house martins to use. Living near a farm with stalled cattle can lead to a hideous number of flies invading the house. So any aerial acrobatic fly-eater was a friend of mine, and any help I could give those friends, I most definitely would.

And so, in the driest parts of the spring, at peak nest-building time, I would creep out at night and line our roadside pond with a bucket of water. As a result, this small ad hoc pond never ran dry, the house martins never wanted for building materials and a rather bemused farmer would often be seen in the periods of extended drought pondering how this short stretch of road managed to magically retain water. As a consequence of this availability of materials on their doorstep, the house martins could get on with the business of building nests earlier in the year. They then routinely produced two and sometimes three broods a year, and in the last three years at the cottage, a colony in excess of a hundred birds could be seen fly-catching in the skies about our house. Clearly, the availability of water on these otherwise arid chalk downlands in the height of summer was a deal breaker, especially for the swallow.

THE *Oxna Mere* is located on Milk Hill, which, like Buttermere, is a place name indicating that cattle were grazed on these hills to produce dairy products for the valley communities below. Oxen, being castrated bulls, are reared for their use as farm labour and, as detailed earlier in this chapter, I strongly suspect that this pond takes its name from the regular puddling of the chalk lining carried out by a team of oxen. So there is the suggestion of a working pond and an indication of how its capacity for water retention was maintained. For me, there is a beautiful intelligence in the *Oxna Mere* reflecting a system of husbandry at one with the landscape around it.

In the criticism of the Hubbard brothers, in the unwillingness to project these ponds back into ancient history, and in the early-twentieth-century desire to scientifically disprove these ponds as having any vapour-capturing capabilities, I believe we have been thinking about them in entirely the wrong way. The scientific endeavours of Edward Martin, as valuable as they have been, are a classic reflection of how the self-confident late-Victorian formal knowledge approach to an issue inevitably leads to an inability to entertain alternative ways of thinking and doing – or alternative forms of knowledge.

In the first instance, there is a failure to consider the ethnographic aspects of these ponds, to consider that they were a part of complex socio-economic systems of maintenance and management. The employment of Portland cement in pond construction, increasingly available from the 1850s, says a lot about the attitudes of this time, that these facilities could only be considered to work if they were deemed to be permanent. Authors such as Edward Martin and the Hubbards were always keen to point to times when folk memory records how ponds have been known to last for sixty or seventy years – that there is merit in their longevity. But there is a danger here in back-projecting a preoccupation with the notion of permanence – doing something once and it lasting for ever. As can be seen from thatching, prior to the widespread use of tiles and slates (with a perceived permanence), the maintenance and management of a roof was considered a cyclic process – factored into the cycle of life. In this paradigm of making, maintenance and management we might entertain the idea that ponds were under a constant cycle of maintenance and management on the downs, and perhaps the association of the oxen with the *Oxna Mere* came about because of the frequency with which this pond was subjected to maintenance, exposed as it was on the very highest, driest and most windswept point on Milk Hill.

In many ways concrete also presented farmers and landowners with a false dawn: it was a material that promised so much but essentially failed to deliver. Puddled chalk and clay will crack under dry conditions, but it has a marvellous capacity, when resaturated, to swell and close up. So not only can it heal itself but it can be repuddled too. When concrete cracks it is permanent. Nothing short of the incredibly costly and time-consuming removal and reconcreting of the pond will suffice. In dismissing in quite such an out-of-hand fashion the capacity of ponds to trap aqueous vapour, commentators in the early part of the twentieth century often failed to consider the locations of ponds for the trapping of rainwater. Put simply, if rainwater is such a consideration, why set them in the highest places? And perhaps again we do our Anglo-Saxon or Neolithic forebears another disservice. They didn't have thermometers and precipitation gauges, but they may well have understood the dangers of too much groundwater run-off.

As my experiences of New Pond demonstrated to me, catching rainwater, and the silts, weeds and fertiliser it brings in, can have a detrimental effect on a pond. Better, surely, for the sake of purer and cleaner water, to keep the pond out of trouble. There is also a biological reason for not catching too much rainwater run-off. In an age before the chemical medicinal treatment of livestock, an infestation of worms could wipe out whole droves if too much manure, carrying the freshly laid eggs of parasites, were to find its way into the ponds. In excavating them into the summit of the hill, it may have been the intention to keep them out of trouble and ensure that the water was the purest it could be. What is so intelligent about these ponds is that you don't need miles of polyurethane pipes, galvanised steel water troughs, self-feeding cistern mechanisms and electric or wind-powered pumps to keep your livestock watered. You just need a spade and a herd of willing oxen.

12

FIRE AND EARTH

COPPEROPOLIS DOESN'T REALLY exist any more. But when it did, from the early part of the nineteenth century through to the first decades of the twentieth, it was the centre of global copper production. Without copper, the British Navy, with its copper-bottomed vessels, would never have ruled the waves in quite the way it did, setting the course for British Imperial supremacy. In a more progressive sense, copper from Copperopolis effectively wired the world for the electrical revolution. Situated on the River Tawe in Swansea, South Wales, all that remains of this once magnificent industry is a handful of ruinous buildings, two tall chimneys and rows and rows of copper-workers' cottages.

It was the former glory of this site that we wanted to recapture when we agreed to meet, in the shadow of the last surviving chimneys, on a cold November morning. I was facilitating a meeting between delegates from Swansea University and a chap called Colin Richards. The objective of our gathering was to explore the possibility of bringing an international smelting competition to the site, to once again breathe fire into the Swansea valley and raise awareness among younger members of the community as to why this part of their city was so important to the Industrial Revolution. Colin is a man of many talents, a virtuoso of any job that involves digging stuff out of the ground and roasting it at high temperatures; he also runs a lucrative sideline as an independent advisor for a wide variety of clients ranging from European conservation bodies to the broadcast media. As I explained to the other delegates what it was he did – iron smelting, brickmaking, tile firing, charcoal making and lime burning – it was Colin himself who most aptly summed up this part of his curriculum vitae: 'It's fire and earth, Alex, fire and earth.'

I like this phrase as it also sums up an enormous number of technological processes that have been instrumental in the development of human societies. There's the full range of metalworking industries of copper, iron, tin and lead, and a whole cycle of processes including smelting, forging and casting that broadly fit into this category: basically, mineral ores extracted from the ground and subjected to fire. Glass manufacture, likewise, can be seen as a fire and earth technology. There are the brick and tile making industries too. These all involve digging up materials, moulding them and subjecting them to intense heat such that their chemical composition is permanently altered. In this sense, they are similar to the ancient craft of pottery – the fabrication from clay of a rich variety of vessel forms to support domestic and commercial economies the world over. In pottery alone the variation is remarkable, ranging from the mundane and everyday to the deluxe and prestigious. Yet, in the spectrum of fire and earth technologies, if the finest ceramic vessels sit at one end, it's lime burning, the most basic and crudest of processes, that sits at the other.

THERE IS NOT the room in these pages to do justice to the high points of the global potting tradition: the T'ang and Sung periods in China, the best of the Ming dynasty, the early Persian and Syrian traditions, the Hispano-Moresque, early Japanese tea masters' wares, the delft and fine slipwares of England, the Korean Ri-cho and celadons, the German Bellarmines. My story of pottery will have to be a much simpler one born from my own experiences, not as a potter but as an archaeologist or 'finder of pots'. One of the virtues of mankind's enduring relationship with ceramics is that they have helped us enormously to identify, date and interpret archaeological sites. Once fired, pots are essentially indestructible. They may break up, heavily abrade and find themselves dispersed across wide areas, but they usually keep enough of their form to allow us to positively identify the period when they were made, and most often, the place where they were produced.

As a British archaeologist, my experiences are derived primarily from sites excavated in southern England, where I became familiar with pottery types from most periods. But the pot-based story that interests me in particular is the one from the middle centuries of the first millennium AD to around the twelfth century and beyond. Here, the pottery evidence has rather casually been used to reinforce the traditional popular narrative of the fall of Rome, the end of civilisation and the plunging of Britain into the dark, barbaric age of the early Anglo-Saxons. It all seems to tie up quite neatly: the Romans were civilised, and they made great pots. The early Anglo-Saxons were uncivilised, and made crap pots. As they became more civilised (by reintroducing classical ideals and converting to Christianity) they started to make better pots. But if we study this whole episode more critically we are forced to reconsider the linear trajectory of progressive technological development – one associated with an oversimplified narrative of decline, fall and re-emergence. It's also a period that is a useful vehicle for exploring what actually makes for a good *cræft* pot and how we go about judging it.

Pottery is one of those crafts where I wrestle with the point at which I see it moving from craft to industry. Like weaving, it's so basic, so fundamental, that certain mechanical and industrial processes were introduced very early on. Take the Romans, for example. They were master potters, but they were also industrialists who, through the use of moulds, made the first steps towards repetitive processes and identikit production. To me, this detracts from the elements of skill so integral to craft production – the hand–eye co-ordination that creates something beautiful in a process where, fundamentally, no two pots can be the same.

By the late eighteenth century, industrial means of pottery production in Britain were well established in Staffordshire, in several major provincial towns and in London. Famous names like Wedgwood, Spode and Minton have their origins in this period and capitalised on a number of developments that were to have a radical impact on pottery production, creating, in essence, a stark break from a tradition of pot production that was at least eight hundred years old.

In the first instance, there was improved access to primary clays, brought about by the construction of a national canal network. Primary clays are hard mineral clays that need to be mined from solid geological deposits found most abundantly in the south-west of England. Known as kaolins, after the Chinese word Gaoling, referring to an area in China where this fine clay mineral was first extracted, these deposits are ground down to make a white clay that fires well under high temperatures. They have a low shrinking capacity, allow for the making of very fine wares and are most famously used in the production of porcelain.

Secondary clays are those that have already been subjected to processes of wind and water erosion and deposition and are therefore already malleable. For the firing of primary clays, temperatures of up to 1,450 degrees centigrade are possible, but the fuel needed to generate this heat meant that such clay was invariably too expensive for traditional country potters. It was the type of fuel that facilitated the growth of industrial potteries, which, like those of Stoke in Staffordshire, were keyed into a network of coal distribution. In this sense, they had freed themselves from many of the traditional constraints placed on local and regional potters.

It is local and regional character that defines traditional pottery, and it is here where its true *cræft* lies. In the first place, secondary clay was as difficult to transport as it was to dig. I know this because I've dug it enough times. As an archaeologist, I loved excavations on the chalk downlands where well-drained, loose and friable soils made life easy. Recovering artefacts from these deposits was like taking candy from a baby. But every now and again I'd be deployed on a site whose underlying geology was a thick clay. My shoulders would slump at the prospect. At best, clay is heavy and sticky but it can also bake hard in the hot sun and be almost unworkable in the rain. So the most important factor in the siting of one's craft pottery was proximity to the raw material. Secondary to this was the need to have access to supplies of fuel and water.

Transportation issues also affected the distribution of the end product. By their nature, the canals allowed for substantial cargoes to glide over long distances into the hearts of manufacturing cities where industrial potters found new markets. The roads of medieval and early modern Britain afforded no such comfort, and the lower-fired wares of the country potter, with less strength and more brittleness, were restricted to the most local of markets. But this had the advantage of stimulating the country potter to be incredibly versatile in catering for a range of needs. The local potter couldn't specialise and expect to survive. Instead, they had to turn their hand to a huge range of vessels from pie dishes, pancheons, cream-making pans, bread crocks, butter pots, stew dishes, casseroles, cauldrons, fish dishes and bakers, to storage vessels, ham pans, salt kits, jelly moulds, jugs, plates, bowls and chamber pots. I could go on. In fact, I will: costrels, spittoons, alembics, paint pots, chicken feeders, hog pots, pitchers, fuddling cups, stinkpots, Long Toms, lading pots, bussas, chafing dishes, bed pans, benisons, barm pots, cloughs, clouts, piggins, posset pots, wash pans, whistles and widebottoms. In many ways, it is this versatility that enabled the country potters to continue their craft well into the twentieth century when, set against the rising tide of mass production, they continued to meet the needs of very local traditions, diversifying into the growing trend for garden pots of all shapes and sizes.

No matter how clean your clay beds were, though, you couldn't just dig it out of the ground and start fashioning vessels from it. There were a number of processes you had to put your local clay through before you could even begin to start thinking about making pots out of it. First, clay needs to be dug in the autumn before it gets too wet, it then needs to be allowed to 'weather', left out in the open to let the frost begin the process of breaking it down. It then requires 'pugging'. This is a form of milling where the raw clay is mixed with large quantities of water to create a liquid slurry. At this point, the potter can choose to add other ingredients to the pugging pit.

They may, as many later potteries did, choose to add a percentage of primary or china clay to harden the end pot and to lift the colour. They may also wish to add a flux material – a mineral oxide that helps to lower the high melting point of the ceramic-forming constituents within the clay, cementing the crystalline components together. This was also the point at which a temper would be added. Temper is made from a whole range of substances, such as finely ground flints, crushed shell, grit or even ground-down pottery, and is added to the mix because untempered clay shrinks to such an extent in the firing process that it becomes unstable. The pugging process takes around twenty-four hours, after which the mix is sieved in order to get any stones or solid matter out before being allowed to drain.

It's at this point that the clay is divided up into balls ready to be thrown, but it still needs to be wedged to drive out any residual air bubbles or pockets of water. Wedging is a technique very much like kneading bread dough, but each ball is divided in half with one half slammed down onto the potter's slab while the other half is slammed down on top of it. They are then kneaded together and the process repeated until the potter – or more correctly, the potter's boy – is happy that the ball is devoid of any air holes. The term 'throwing pots' derives from the forceful manner in which the ball is thrown onto the potter's wheel, and it's from this point on that the mesmerising and manipulating caresses of the master potter take over as the vessel form takes shape.

The hands must be kept wet at all times as they shape, raise and flatten the ball. In response, the clay bulges, swells, contracts and opens up as fingers and thumb begin to thin the walls. One hand counteracts the other as the emerging form is tended in an episode of perfect hand–eye co-ordination, before being finally tested for size with the gauge and removed from the wheel with a cheese wire. It is the hands, the subtlest of all machines, that are the crucial tools in this process, rather than a reliance on any fancy equipment; they must determine the balance between volume of material and strength. This is a knowledge that only the hands possess.

The skill of potting is not over with the act of throwing. These 'green' pots need drying in a measured and considered way – traditionally, in conditions that varied according to the prevailing weather. Although drying rooms are used by many large-scale enterprises today, as most craft potters will tell you, any acceleration of the drying process impacts on the quality of the end product. It's always better to use natural drying conditions, to allow the material to settle at its own pace. And it's no good being the best thrower of pots in the land if you can't match that skill with a knowledge of firing. In the first instance, you need to know when the pot is ready to go. You need to understand how the pot is going to behave when subjected to extreme heat in the firing process – and by how much it will shrink. You have to get the stacking of the kiln right, understand its thermodynamics and allow the correct cooling time. Small windows into the kiln, or a removable brick from its body, enable the potter to inspect the process, but *knowing* the success of the burn will come mainly by judging the heat levels from the colour of the pots.

Glazing represents an important development that served both decorative and functional purposes. The first firing of a country potter's kiln will produce what is called earthenware, with a porous, biscuit-like finish. These pots will hold water and can be used for cooking, but through use they will also take water into the body. So, to make vessels water-resistant, they need to be subjected to a second firing, having been coated with a mineral composition made from silica, boric, lime, oxides, potash or salt, but often from lead or tin oxides. This creates a vitreous and glasslike coating that protects the underlying fabric of the pot and results in a range of attractive colours, depending on the conditions of firing. By the thirteenth century, glazing came as standard for most types of pottery, and in the later centuries of the medieval period this tradition came to be known as majolica or faience – a tin glazing of earthenware that was still used in the production of fine delicate delftwares in the seventeenth century.

By the mid-nineteenth century, this had largely been the process of pot manufacture among the country potteries of Britain for the best part of eight hundred years. The situation was not entirely static: staffing structures, markets, fashions, fuels and clays will all have changed slightly. Gradually, areas of better clay and fuel supported growing regional production centres at the expense of very local potteries, as improvements in road surfacing and canalisation of rivers developed communications from the late medieval to the early industrial period. Scale may have changed, but the principles of making a pot from local clay, using local fuels for local supply chains, were essentially the same as they had been since the twelfth century.

While we can work back from what is known of later potting, tracing the origins of this mode of manufacture is a little harder. The Battle of Hastings in 1066 looms large in the categorisation of many archaeological entities. Pre-Conquest and post-Conquest have become forms of periodisation that when applied often mask the complexity of much longer-term trajectories of change and development. There is no doubt that by the early twelfth century pottery production was a specialised craft industry with dedicated sites producing a range of wares for local consumption. But whether these modes of production were introduced as a direct consequence of the Norman Conquest is another matter. They might just as well have been part of a longer-term technological shift taking place on a pan-European level.

While I've already alluded to the crude wares of the early Anglo-Saxon potters, it should be mentioned that by the tenth century the late Saxon wares of East Anglia, such as Stamford Ware, St Neots-type ware and Thetford-type ware, exhibit a remarkable degree of sophistication. There is even evidence that as the industry at Stamford developed, certain wares were being glazed with lead oxides. Famously, East Anglia was settled by Vikings who were connected into a wider European network of trade and exchange that was to serve as a stimulus for economic growth in England as a whole. So it seems more appropriate to index technological developments in pottery to developments in urbanism and regional wealth generation rather than link them to Norman or Anglo-Saxon political control.

But what exactly *was* the Anglo-Saxon potting tradition before the economic developments of the tenth century and the likely introduction of continental styles of ceramic production? One of the main issues with Anglo-Saxon ceramics is that they are difficult to date because there was little in the way of improvements over the course of the early and middle centuries of the Anglo-Saxon period. The pottery recovered from these early periods is often referred to as chaff-tempered, but has been found to contain a range of organic materials such as grass or dung. It's often crude in its form, not wheel-thrown and is low-fired and burnished which, in being softer and more friable, impacted on its survival in the archaeological record and the ability to spot it in the process of excavation. In the early days of analysis, it was associated with the pagan period, and it's easy to see how early commentators unquestionably worked it into a narrative of technological collapse following the fall of the Roman Empire.

But there is a different way of spinning this story. In the first place, the Anglo-Saxons in rural settings may simply have been less reliant on pottery than earlier and later cultures. This doesn't make them less sophisticated. Their ability to turn wood and craft leather could easily have furnished them with substitute vessels. That evidence for this is short on the ground is because such organic materials rarely survive. Perhaps the old adage 'If it ain't broke, don't fix it' applies here, and the reason organic-tempered pots are found across numerous centuries, from the fourth to the ninth, is because they were effective – if not as effective as later pots to use, then at least more effective to make. It's the technology that matters here. With materials like grass, chaff and dung being used as a temper, it's easy to assume that these cheaper materials produced substandard goods. In fact, the voids created by using such materials can reduce the thermal shock properties of low-firing temperatures. In the fabrication process too, organic tempers can improve the workability of what is called short clay – clay that lacks plasticity – which is often the case when using super-local varieties. Despite its perceived inferior quality, the technology in its production was highly suited to domestic craftsmanship among mobile agricultural communities.

If I admire the *cræft* of the country potter – using local reserves of clay and fuel to serve local markets – then in many ways I find the resourcefulness of the Anglo-Saxon domestic potter even more appealing. It's a lesson for us all. Studies like this force us to consider some of the big questions in societal development surrounding sustainability and resilience. Despite the sophistication of Romano-British pottery, what happened when the networks of clay extraction, fuel requisition, manufacture, transport and marketing imploded? What, then, did people have recourse to, having placed their dependence on industries reliant on macroeconomic structures that were ultimately beyond their control?

L IMESTONE IS CALCIUM carbonate stone, and when you roast it at a temperature of around 900 degrees centigrade a chemical reaction takes place and molecules of carbon dioxide are liberated. This process is known as calcining or lime burning, and it leaves you with calcium oxide, a solid also known as quicklime. Quicklime is unstable, which means it spontaneously reacts with the carbon dioxide in the atmosphere in a bid to return to its original dormant state. This is a relatively slow process and can be arrested by slaking – submerging the calcium oxide in water. The problem is, during the process of rehydrating, quicklime goes through an exothermic reaction, meaning it gets hot when it comes into contact with water – very hot. It's so hot that, although in itself it's not a combustible substance, it can spontaneously ignite any organic material it comes into close contact with. This can be a problem, for there is moisture in your eyes and mouth, nasal cavity and lungs – so breathing in its powder form is highly dangerous. In fact, so dangerous is it that quicklime dust is recorded as a weapon of war in ancient and medieval times.

For the most part, however, the uses of lime have been of a more benevolent variety. While it's volatile and incredibly caustic, its desire to return to its rehydrated calcium carbonate form means that it sets back almost as hard as the stone it once was and, as a consequence, has significant bonding properties. As you can imagine, it has been popular for all manner of craft and industrial endeavours over the centuries. Indeed, its use may represent the first evidence we have of human beings purposely constructing their own accommodation. Archaeological excavation of early Neolithic sites has recovered evidence for layers of lime mixed with clay in what look to be primitive floor linings. So human beings may have been making solid plaster floors before they were even firing pots.

Over the course of history, it would appear that lime's primary use was as a building material, in the bonding together of stone and bricks to make walls but also in the rendering and plastering of them. However, in the nineteenth century, a newer, more robust material was developed. While cement – a compound of lime and pozzolans (such as volcanic ash or brick dust) – has been around since Roman times, Portland cement, as it's come to be known, was a more aggressive, quick-setting, harder and more durable product. Nowadays, therefore, the use of lime in building materials, except on historic properties and eco-builds, is a rare occurrence. But lime is far from dead as an industrial material. In fact, we produce more quicklime today than we have in any other period of history – to the tune of hundreds of millions of tonnes globally – chiefly for use as a flux in the steel-making process. But because much less of it is used in regular building work, there is less of the stuff lying around and available for use on a day-to-day basis. We have therefore lost the knowledge of the multitude of functions lime has fulfilled for us over the passing centuries.

My first real introduction to lime was in the excavation of a medieval limekiln in Southampton. I'd secured some short-term archaeological digging work with Southampton City Council's archaeology division. The department itself was a bit of a dinosaur, a throwback to the municipal divisional arrangements of the 1970s. At a time when most council archaeology units had been shut down to make way for commercially subcontracted firms, this ragtag bunch of specialists still plied their trade along the old lines. They weren't the most competitive or efficient outfit in the region, but nobody was more enthusiastically knowledgeable about medieval Southampton. So they were a pleasure to work with, and my first site with them was an area immediately outside the line of the medieval town walls.

Early analysis suggested that, beneath the remnants of a later friary garden, there wouldn't be a huge amount of archaeology – perhaps some 'ephemeral extramural activity' (archaeologist-speak for 'outcast vagabonds living in makeshift huts beyond the pale of medieval civilisation'). And indeed, when the mechanical digger peeled back the upper layers of topsoil, in preparation for a new housing development, the site was largely devoid of any substantial archaeological remains. Scatters of medieval waste, a few pits and numerous garden features were what we spent the majority of the two months excavating and recording. At one point we exposed two parallel lines of what we thought could have been huge post pits. Immediately, visions of a vast Saxon palatial timber hall sprang to my mind. But, sadly, you don't get paid to dream on archaeological sites. Excavations confirmed that each post pit was more likely to have been a tree root bole staining the soil in a circular fashion. That they were in rows indicated they were probably part of an orchard – a feature entirely consistent with a friary garden.

As is often the case on this kind of project, it's towards the very end of the excavation, when you've finally got your head around the nature of the archaeology and the story of the site, that a few last-minute surprises emerge. In this instance, it was a strange circular feature with numerous halos of fired clay that was exposed in the cleaning off of the final corner of the site. I was immediately set to work excavating and recording the find. It's on sites like this, where the archaeology is paid for by the developer, that one has to adopt a fine line between meticulous archaeological examination with archival standards of recording and bashing it out with pick and shovel and making a rough sketch in the site notebook. I tried to err on the side of the former approach, but with only a few days before our time allocation was up, I had also to be seen to be working quickly.

In plan form, the feature was circular, but closer inspection revealed a flue-like entrance to the north. The halos of burning were evident in the scorching of the subsoil. Here in Southampton, the underlying geology is a heavy clay known as brickearth, by virtue of the fact that it makes excellent bricks. When you fire clay at different temperatures, it turns different shades on a spectrum from light yellow through to deep purple. When you fire clay *in situ*, that is to say, unexcavated, the clay closest to the source of heat turns the darker colour with the spectrum running to the lighter shades radiating away from the heat. The circular feature itself was a pit lined with stones, and as I cut a section through it I exposed layer upon layer of charcoal interspersed with layers of a fine white powder – lime. There was probably only a third of it left, with the top two-thirds being levelled off for the setting out of the garden. The medieval lime burners may very well have chosen this location because of the consistency of the underlying clay and its suitability for ad hoc kiln construction.

But what were they using the lime for? As I finished cleaning up the excavated kiln, photographing it, completing the section drawings and filling in the various recording sheets, I looked around me. Southampton was once a famous medieval city. In its prime, it had matched London for its maritime economy, with trade from France, the Atlantic seaboard and the Mediterranean beyond. Goods landed here could just as easily be sent to London via road rather than be circumnavigated around the Sussex and Kent coastlines and up the Thames. Southampton flourished as a consequence and enjoyed a period of civic pride which is best illustrated in the wealth and power invested in the construction of overtly symbolic and ostentatious town walls. Sadly, during the Blitz, these walls, along with the rest of medieval Southampton, didn't fare too well and only a very small proportion of what would undoubtedly have been a stunning medieval citadel has been left to us.

But overlooking our archaeological site was a remnant of a bastion, a semi-circular tower that would have fortified this particular stretch of the town wall. After work that day, as the mechanical diggers moved in to destroy what was left of my kiln, I wandered over to take a closer look at this remnant of Southampton's medieval past. Stone built, it had stood the test of time and in later periods of relative peace had been reused by the friary as a dovecote. With small nesting holes cut at regular intervals into its internal walls, the bastion was still proving popular with the city pigeons. I examined in fine detail the mortar that had bonded the cut stone together for over seven hundred years and observed tiny flecks of charcoal and fired clay. It could never be proven, but it's not impossible that this mortar had been mixed from lime fired in my makeshift limekiln cut into the malleable clay of the Solent basin.

IT WASN'T FOR another ten years that I got the opportunity to fire a limekiln. We were making a television series for the BBC, and lime production was an important industry for the site we were using as our main location. This was evidenced by two enormous limekilns alongside the tidal river, the perfect position for the bringing in by barge of fuel and raw materials. The moment my colleague and fellow presenter Peter and I clapped our eyes on these kilns we became obsessed with using them to see if we could roast limestone and turn it into lime.

After almost incessantly nagging the production company, we were given permission to at least measure up the kilns. Peter skipped off to get an old measuring tape while I sought out a ladder to allow us entrance into the depths of the kiln. Unlike my kiln in medieval Southampton, these were on an altogether different scale. They took the form of giant funnels cut into the cliff side. At about twenty-six feet deep and twenty feet across, we calculated a capacity of around 4,200 cubic feet. As Peter paced out the circumference, dropped the tape down into the kiln and measured the diameter of the base, my heart began to sink.

There's a ratio of fuel to stone that goes something like two to one. I quickly realised that to fill this kiln and operate it, as it would have been in 1900, would be prohibitively expensive for the production company for a scene that would deliver a mere few minutes of television. And soon other problems loomed on the horizon. First, in the burning of limestone, a series of toxic gases are given off, and these kilns were in dangerous proximity to sites of human habitation. Second, the handling of the resultant lime was not going to be without difficulties. Quicklime is a highly volatile and combustible material. We'd need to hire some pretty officious health and safety executives to draft up a risk assessment for the handling and disposal of twenty pounds of the stuff – but we were intending to produce around thirty tonnes' worth. The third and arguably the most boring reason is that these were historic buildings. They were being preserved, and it was considered that burning sixty tonnes of coal in them, at a temperature of around 1,200 degrees centigrade, might have a detrimental effect on their long-term survival.

I (sort of) understood this mentality but I railed against it. Preservation for me is as much about raising awareness of our heritage and thereby encouraging people to take part in its protection, both financially and in person. But I also conceded that the fabric of this kiln was tired and fragile. The firebricks that lined the cone were in pretty good shape, but their surfaces were friable and they had clearly taken in moisture. Striking a fire of extreme heat represented a potential threat to the overall structural soundness of the entire monument. As the moisture expanded at speed it would very likely shatter the surface of the brick lining, doing permanent damage to the whole structure.

I understood the concerns of the limekiln preservationists and it appeared that, for now, our lime-burning ambitions had to be put on hold. But a few months later the prospect re-emerged as a distinct possibility, courtesy of Colin Richards. With his MBE for services to historic building conservation, he was well placed to advise on a range of our proposed projects, having already organised the firing of a brick kiln for an earlier BBC series. But Colin wasn't the sort of chap to simply advise from behind the comfort of a clipboard; he was very much the kind of consultant who rolled up his sleeves and mucked in. Mild-mannered, considerate and genuinely enthused by our ambitions, he had located a recently refurbished limekiln in his home county of Shropshire. Permissions were acquired for us to use it – under Colin's supervision – and because the kiln was so remote, there was no issue with intoxicating local people as they slept in their beds.

But one problem still remained – the sheer volume of the kiln. On a slightly more moderate scale than originally planned, we were still going to need twenty tonnes of coal and ten tonnes of limestone to produce around five tonnes of quicklime. And this stuff, as everyone kept telling us, was incredibly dangerous – especially when it came into contact with water. One expert informed us that if an amount the size of a pea were to land in your eye, in reacting with the moisture it would completely burn out the socket before it achieved a dormant state. It never took much to frighten the desk-based health and safety officers at the production company in London, and tales like this did little to allay their fears.

We proposed the most sensible personal protective equipment along with assurances that we would proceed with the utmost care, but still their anxieties triumphed. Never underestimate the capacity of a risk assessor to confuse deep water with drowning. Deadlock had been reached. Could we at least scale it down? came the message from head office. Not really. There was, local expertise argued, a point of critical mass required to make the burn effective. That is, you had to have enough material in there to reach the requisite heat to roast the limestone to the point of chemical reaction. I considered the modest scale of the medieval limekiln I'd excavated in Southampton. It obviously worked on that scale, because the site was subjected to repeated burns. But I kept my mouth shut. I was holding out for a big burn-up.

In the end, the case for using this new kiln to maximum capacity won out. We argued that it wasn't just the filming of the lime-burn sequence we could benefit from, but we'd done a bit of research and realised that there were a near infinite number of subsequent scenes we could utilise it for, making the whole endeavour financially viable.

So, with firewood, coal and limestone ordered, we set about stacking the kiln. This was a tricky business. First, we needed to stack the firewood at the base to get the burn going. This would be covered with a layer of coal then a layer of limestone. We needed to make sure that each layer was put in evenly, both in terms of the ratios of coal to limestone and the way it was spread about. The initial plan was to have someone in the base of the kiln spreading it out as the other members of the team shovelled it in. Peter drew the short straw on this one and, with rake in hand, he duly made his way down the rickety old ladder. Of course, this plan worked for a matter of seconds. As the first few shovel-loads of limestone came raining down, it became obvious that we'd more than likely stone the poor chap to death before getting anywhere near filling the damned thing. With Peter winning a reprieve from his kiln base duties, we determined that to keep an even distribution of materials we would count out the shovel-loads as they went in. Four men, the sound of shovels scraping, the dashing of coal on limestone and limestone on coal, and the relentless monotony of counting out each and every shovel-load. It was nearly midnight by the time we finished. But at least we were ready to go.

Starting the inferno wasn't easy. This was done through a stokehole entrance located in a small vaulted chamber external to, but level with, the base of the kiln. Being damp and cold, it was difficult to get a draw, but after the best part of a box of matches and a considerable amount of wafting with a sheet of tin, the timber in the kiln base started to crackle. And so the wait began. All in all, the kiln took about three and a half days to burn out. You might think that there is little to do once the fire is lit, and it's true there aren't a great many moving parts on a limekiln to tinker with. But the crucial thing is to ensure that there isn't too much of a draught blowing into the stokehole. You do this by damping – placing a board on or against the entrance. Obviously, you need to get oxygen into the fire in order for it to burn, but if too much air races in through the stokehole it can create a column of burning in only one part of the kiln, which in turn has a kind of chimney effect. The fire races up this flue of extreme heat, charged as it is by the blast of air, and burns out, having bypassed much of the stone and coal. Back in the nineteenth century, this would have resulted in a kiln load of unsellable, half-burned limestone – and the end of your lime-burning career. But damp too much, and there was a risk that the fire could be put out altogether.

We had a tricky first twenty-four hours as a south-westerly wind blew hard at the kiln face, but it died down soon enough and we watched and waited as the contents of the kiln bowl slowly diminished in a mesmerising glow. We took it in turns to stand guard, using the small vaulted chamber as a base from which to keep warm and keep an eye out. Occasional visits were made to the top of the kiln to stare down on the hellfire and brimstone-like cauldron we'd thrown the quarried stone and coal into. Like a witch's brew, noxious gases swirled out of the basin and a strange glow radiated upwards. It felt distinctly like we were conjuring up a very ancient form of alchemy. Transfixed as we'd become, we were careful not to spend too much time on the precipice. We'd heard stories of how wandering vagrants, in seeking warmth in the harsher winter months, had curled up alongside the edge of a kiln, succumbing to the toxic gases as they slept, only to be found stone dead the next day. Our visits to the summit were to check that the burn was even and that the volume of material was diminishing evenly. From a small peephole located at head height in the vaulted chamber we were able to observe the colour at the heart of the burn: shades of orange, deep red and purple as the heat of the fire shuffled its way through the contents of the kiln.

After three and a half days of us camping out, the burn had finally extinguished itself, the structure of the kiln had cooled and five tonnes of quicklime was ready to be removed – by hand. Strangely, our fellow crew members and lime-burn enthusiasts seemed to have found more important things to do by that time ... but we weren't too bothered, for the material we'd created – under the masterful auspices of Colin Richards – was of an astounding quality. As a test, we dropped a lump the size of a tennis ball into a large, deep puddle of water. The quality of quicklime can be measured by the scale of the chemical reaction it has with water and, believe me, this thing went off like a banger and fizzed and skimmed around the puddle for a good few minutes. We grinned at each other but quickly realised the inherent danger that presented itself in the handling of over five tonnes of this stuff.

We spent the best part of seven hours carefully shovelling it into airtight oil drums, taking regular breaks to keep focused on the job in hand. And so much for our fears about finding somewhere to dispose of it safely. Word soon got out that there was some good stuff to be had down at the limekiln. Builders, renovators, conservationists and plasterers came out of the woodwork. Commercial lime was readily available enough, but we had the real deal, as pure and authentic as lime gets, and people were desperate to get their hands on it. Peter and I became illicit dealers of fine white powder, by the hundredweight. In the end, we managed to offload the bulk of it and took three drums back to our location for use in a range of building projects.

I T IS AS a building material that I've become so impressed with lime. Its demise for this purpose has largely been put down to the introduction of cement, and in a straight fight between the two, especially for the needs of twentieth-century urban development, cement has increasingly come out on top. It sets harder and quicker and is water-resistant. It's a no-brainer. The concrete jungle was born. But only very recently there has been a realisation that cement isn't quite as clever as we thought it was, and for a number of reasons. In the first instance, in its production there is a considerable amount of damage done to the environment. It has no real reuse value either – unlike lime. Amazing as it may sound, lime plaster removed from a wall can be ground down to a powder, rehydrated, mixed and replastered. It's a labour-intensive process, but I've seen it done and it really does work. Concrete, on the other hand, having set hard, is in that fixed state permanently. And the force required to dismantle and process it into a basic hardcore is almost as energy-hungry as producing it in the first place.

One of cement's strengths is also its weakness. Any substance that can set underwater, is insoluble and water-resistant is therefore also impenetrable to water. When building a house, you might think this is a good idea. You don't want water driving through the walls and causing damp. And so concrete works as a render to achieve this purpose. But what if you get a build-up of water in your house? In which case, it has no way to get out, remains trapped and can cause damp rot, mould and an ideal environment for all sorts of unsavoury fungi. You'd be amazed at the amount of water that can build up in a house. Every time you boil the kettle or have a hot shower or bath, plumes of steam rise and condense on the walls and ceilings. And just breathing alone can produce airborne moisture that can condense against cold and already damp walls. These aren't a problem if you have bottomless pockets, because you can just keep the heating on, crack the windows open and cook the house off from the inside out. But if you're on the breadline of heat poverty, plain frugal or simply more conscious of the environment, the limited heat in the building fails to remove the moisture. The solution, of course, is to open the windows regularly and allow the natural air to circulate and lift the moisture away. You don't need warmth to dry things off, just dry and windy conditions. But if you work away from home all day this can present a security issue, and at colder times of year it's hard to keep yourself warm inside a house with all the windows open.

Certainly, when fuel costs were lower, cement seemed like a good idea. But now that we're entering a phase of fuel economy, the impermeable properties of cement don't feel too clever. Lime, on the other hand, is a breathable material: air and water can pass through it. It's like a Gore-Tex lining, the original breathable membrane. It can also move. It sets hard, but not so hard that it can't flex to endure long-term structural pressure or the swelling and shrinking caused by heat and frost and their impact on moisture retention. If water gets behind a concrete render and freezes, the expansion will cause the cement to fracture. These cracks will allow more water in, water that becomes trapped because the concrete can't breathe – and you have an exacerbating problem that can spiral out of control. If lime render gets wet and freezes it swells, but as it dries and warms it shrinks again. It's more of a living and less inert substance than concrete.

To MAKE A limewash, one only has to water down a lime plaster to a thin milky soup. This can be splashed on liberally to spruce up a tired internal wall or, with a thicker mix, worked into an external wall to pull the weathered surface back to life. Over the centuries, lime has been liberally applied to buildings of various stature and was responsible for many an iconic regional building. Take the authentic Tudor look, for example: black timber framing infilled with panels of white plaster. Tar or pitch was used to weatherproof the wood, while a generous coating of lime plaster and wash protected the soft clay daub beneath. But even stone-built buildings were lime plastered and finished in a limewash. This had a dual function. A coating of lime render can aid run-off and avert damp penetrating the mortar layers between the stone. But it can also give your building a uniformity and beauty that in some instances is useful – especially if you want your important building to stand out.

The place name Whitchurch (there are examples in Hampshire, Somerset and Shropshire) is an indicator that lime rendering was practised in the early medieval period on the most important building of the community. Only the other day, I was wandering around an old farmstead in the Highlands of Scotland and noticed from the stonework that originally this building was intended to have been rendered in lime. The dressed stone around the windows, doors and quoins (cornerstones) stood an inch and a half proud of the walls, providing a hard edge against which to render. Below this the wall was comprised mainly of granite – a stone harder than any north-easterly wind, no matter how foul. So, you might ponder on why the need to provide a protective outer layer. In the first instance, the courses of granite stone were uneven, staggered and in places infilled with small irregular limestone blocks – so it all looked a bit messy. Although now standing in a state of decay, this farmstead, dating to the 1870s, was once a spectacularly modern construction, forward looking and conceived in an age when grandeur in farm buildings was a means of showing off wealth and status. Any proud landlord would want his new investment – in this case, a dairy establishment – to stand out from its natural stone surroundings. And lime was the perfect material to do this.

But lime on buildings also has another function, and this was illustrated by the rendering of lime that would have continued from the external walls through into the walls of the courtyard of this fine range of farm buildings. When it comes to sanitation, there are few substances more naturally adept at deterring disease and pestilence than lime. In the context of this courtyard, worn tethering rings fitted to the walls suggested that cattle were stalled on a regular basis. Dung, milk and mucus – potential carriers of mites, bacteria and various diseases – would have splattered against these walls with, for the dairyman at least, an irritating frequency. Lime was the means to arrest infection and disease on the farmstead in the age before chemically enhanced detergents because, even after slaking, it remains a pretty caustic substance. Using it with bare hands takes a considerable toll as it soon dries the skin surface causing cracks and bleeding, and I imagine it has the same desiccating result when in contact with insects, fungi and bacteria.

You might have heard the urban myth about how drawing a

chalk line across an ant trail dissuades the ants from persisting in their course. It sort of works, but only for a short period, then they figure out what's going on, realise it's only a thin barrier between them and their desired pathway and they cross the line to continue about their business. But there is a ring of truth in the myth. Chalk is a form of limestone and thus contains calcium carbonate. In fact, chalk makes a pretty good lime powder, and a barrier of chalk-derived lime definitely stops the ants in their tracks. This is especially useful for timber buildings since an infestation of hungry ants can do almost irreparable damage.

MY OWN ADVENTURES with lime began with a round of pest control, which was much required during a year of heavy breeding within my chicken concern. I'm a bit of a fanatic when it comes to chickens. Every autumn I tell myself that four or five hens and a cock-bird to look over them as they feed is all that my small family requires to keep us in fresh eggs throughout the year. And yet, as winter turns to spring and my hens ramp up the laying and eventually turn broody, I entertain visions of a bumper crop of chicks, the breeding of my own strain of pedigree hens, and a cash crop of pullets to sell at the end of season poultry fair. It's not that I need the money, it's that I like to get my poultry concern to pay for its own upkeep. When the rest of the world is telling us to think big, never underestimate my capacity to think small-time. Ultimately, when you factor in the time taken to care for your chickens, and the cost of feed and housing, the cheapest way to acquire eggs is to pop down to the local shop and part with £1.60 for half a dozen. But that's not the point. I wanted free-range eggs produced by my own chickens feeding on the bugs and seeds in my garden.

It was during a year when I was looking to specialise in the Light Sussex breed that I ran into problems. The Light Sussex is a great breed. It's one of the oldest recorded in England and known as a dual-purpose bird – good at laying eggs and pretty good as meat. They'd proved popular in the end-of-season poultry sale the previous year, and I thought they were a safe bet for a reasonable return on investment for the coming season. Things went well. I'd stored up a respectable clutch of eggs over a couple of weeks from the hens intended to produce chicks. As other hens went broody, I carefully replaced their own eggs with my selected ones and waited for hatching. By late June, my poultry concern had already grown from eight birds to thirty-eight. I'd learned very quickly to sex the chicks just by sight – by the way they stand – and the early signs were that at least 50 per cent of my new generation was female. I did some rough calculations. At a minimum of fifteen pounds per bird, fifteen birds would bring in £225. My annual feed costs were around £90 and I'd invested in another nesting pen at £125 that year. So, I'd covered my expenses and made a profit of ten pounds. Everything was going to plan.

And then things started to go wrong. As the chicks fledged and began seeking out roosting perches, their modest accommodation (a couple of six foot by four foot sheds modified to take perches and nesting boxes) reached capacity. The young cock-birds also started to fight and disrupt the whole flock. In short, it all got a bit stressful, and stressed hens don't lay eggs. I decided that it was time for some of the cock-birds to go. I wanted to keep two of them to sell on, but the others were very definitely destined for the pot. So that day, I wrung one neck and we had a superb coq au vin that evening. It was absolutely delicious, but much in need of a bit more meat on the bone. The others I locked into a shed with a small external run. This would keep them out of trouble and, with access to as much food as they could eat, would fatten them up nicely.

It was during this short stay of execution that an infestation of mites broke out. Hatching from tiny eggs laid in the nooks and crannies of the sheds, the mites would come out at night and feed on the roosting cock-birds. Unchecked, I knew the cycle of mite reproduction would cause chaos – especially if they spread to my other roosting shed and to the nesting and laying birds. So I had to act. Desperate to find a quick-fix method, I perused the shelves of the local agricultural supplier but couldn't find any substance that was chemical-free. Creosote, a by-product of the coal-tar industry, was recommended to me, but I was hoping to find something with more environmental credentials.

And then I remembered a couple of buckets of slaked lime that I'd kept back from a plastering job I'd done a while back. By watering down the smooth paste, I mixed up some limewash for the sheds and duly set about applying it. It was great stuff to work with. You could really splosh it about, working it into the cracks and crevices of the shed and coating the perches and floor. I did the same to my other shed. And it worked. Some of the birds' feet looked a bit raw after a couple of days, and in one instance I had to reach for the antiseptic cream. But if anything, it did their feet a world of good – I'd never seen some of the older birds with cleaner feet. There is a useful point to be made here and it lies in the necessary steps one has to take when more intensive methods of production are adopted. The moment restrictions on space and numbers become an issue is the moment you get a build-up of unwanted nasties. Any species that is effectively monocropped in a non-biodiverse environment can create a breeding ground for parasitic insects, bacteria and fungi, and it's at this point that you need to turn your hand to remedial cures. Today, a series of potent chemicals has played a large part in allowing us to sustain intensive production in livestock enterprises, but traditionally limewashing and liming was the most effective means of keeping a lid on infestations.

And it wasn't just around livestock that cleanliness could be maintained through the judicious application of lime. It was found, in the later nineteenth century, that plants could also benefit from being dressed with a lime-based mixture. Most famously, it was the winemakers of central western France and the Bordeaux region who were to benefit from this revelation. A generation earlier, economic conditions and improved communications in an industrialising Europe meant that many more markets were open to producers and sellers of Bordeaux wine. As a viticultural sprawl took hold in the landscape of the region, so too did an infestation of mildew. Appearing as a dark stain on the leaves of the maturing plants, it quickly spreads and destroys the leaf matter entirely, denuding the vine of its ability to produce grapes. This particular mildew was thought to have been accidently introduced from vines brought over from the Americas; the native French vines had no natural resistance, so it spread like wildfire. At one point, the future of French winemaking hung in the balance.

But serendipity played its hand and within a decade things were looking up. The story goes that a grower, tired of seeing his roadside vines stripped of grapes by passers-by, had sprayed the plants with the most noxious solution he could lay his hands on, a mixture of copper sulphate and slaked lime (calcium hydroxide). It just so happened that later that year Pierre-Marie-Alexis Millardet, a botanist whose interests lay in trying to find a cure for the afflicting mildew, was passing and he noticed that the plants sprayed with the scrumping-deterrent mixture had mildew in a much arrested state from those growing further away from the road. Additional trials using different mixes resulted in the widespread adoption of what was to be called Bordeaux Mixture, and right up to the present this substance, considered suitable for organic horticulture, is used in the treatment of grape vines, but also in the treatment of all manner of fungal infections, including potato blight and peach leaf curl.

Lime is also useful to growers as a fertiliser. 'Marling' a field is the process of liberally spreading a calcium-carbonate-rich mudstone or clay across a field to enrich the soil. This practice is hundreds of years old, but it was only relatively recently that we came to understand the benefits of mixing lime-rich conditioners into the fecund earth. In the mid-nineteenth-century it was considered a somewhat crude and ancient process, thought mainly to have been conducted to bulk out thinner soils and give them more heart. But as scientific research developed towards the end of the century, it was established that the beneficial effects were as much chemical as they were physical. Not only did these additives neutralise the acidity of the soil but they could also liberate vital plant nutrients from the soil matrix.

While marl is a dormant substance excavated from the ground, the more aggressive and potent form of it – quicklime – can have an even more invigorating impact on soil quality. As sources of fuel for burning limestone became cheaper and more widely available throughout the industrialisation of Britain, more lime fertiliser could be produced for agricultural purposes, and as a consequence, along with other chemical and technological developments of the age, yields rose considerably. What's interesting is that while the industrial agriculturalists had the science to back up and qualify the expense of top-dressing their fields with lime, the ancient practice of marling is an indicator that much further back into our past there was a knowledge that such practices could improve the quality of the soil.

SINCE THE FIRST days of excavating that limekiln in Southampton – a time when I was really getting my head around exactly what lime was – I've always kept my ears open for different uses of this most versatile of substances. For years I'd never thought twice about the term limelight and how it could be stolen from someone by dint of upstaging them. I realised, of course, in my new lime-enlightened state, that this must be light that is provided by lime. Sure enough, blast oxyhydrogen flames through blowpipes at cylinders of calcium oxide to a heat of around 2,400 degrees centigrade and you can achieve an incredibly bright incandescent glow. It's the perfect way to place the star of the show in glorious spotlight.

But I think the most bizarre use for limewash I've ever come across was the recommendation in a 1914 Ministry of Agriculture and Fisheries manual that advised storing your glut of chicken eggs in tin buckets containing a much watered-down limewash. In spring, when the birds really start laying, eggs come so thick and fast that you end up with too many to eat or sell. So you need to find a way of preserving them for later in the year when the cold and shorter days put most breeds off laying. In an age before refrigeration, a bucket of limewash was a viable option. Eggshells are effectively calcium carbonate, and submerging them in a solution of the same substance protects the precious cargo inside from oxygen. Doing this creates anaerobic conditions where, deprived of oxygen, they pickle and keep for much longer than if they'd been placed in a cold store. I've tried this and it definitely works, though the shells get incredibly brittle and it's difficult, when cracking the eggs, to keep the tiny shards out of the white and yolk.

It's clear that lime has proven itself to be a valiant partner in the story of human endeavour, fashioning its way into all manner of trades and industries. In so many ways it's a chemical that, through its reaction with other substances, brings a beneficial outcome for those using it. But it's also something drawn from the earth, something very basic, whose altered state can be created simply through the equally basic process of fire.

13

THE CRAFT OF DIGGING

WHILE I CAN talk a good craft, I'm no craftsman. I've turned my hand to all sorts of things over the years, and at times of brimming self-confidence I like to consider myself a Renaissance man, but when I get down off my high horse the expression 'Jack of all trades, master of none' is definitely more fitting. But if there is one thing I'm pretty good at, it's digging. By this I don't mean digging ditches – although I've dug a fair few in my time. When I mean digging, I mean archaeological excavation. I've often struggled with the idea that excavation is a craft. Sometimes I see it as a science, a series of methodological steps to deduce a series of processes – past events – from a set of physical remains. A bit like forensic science. At other times, I think it's more about basic practical aptitude. So, in the same way that you can either put up a set of shelves or you can't, you can either dig or you can't. But then at other times I'm seduced by the idea that excavating requires some kind of ineffable ability. When fellow archaeologists used to say of someone else on the circuit that they could or couldn't dig, the insinuation was clear: anyone can dig. But can you *dig*?

Digging has obsessed me from a very early age, and I assume this is one of the reasons why I gravitated towards archaeology. I remember marvelling as a little boy at how quickly my father could dig. When I grew up, I promised myself I would be able to dig as fast as my dad. For one job, when he wanted to run a service trench down the length of the front drive, he considered hiring a mini-digger and driver. The quote, when it came in, seemed pretty reasonable. But being the tight-fisted Scotsman that he is, my dad had other ideas. Emerging from the shed a few hours later with his self-made hybrid spade – a scaffold pole hafted to a steel spade head – he set about digging the trench himself. It was how he used it that I was so intrigued by: thrusting, kicking, levering and flipping in a fluid and repetitive motion. There was clearly something of the Highlander in my father's blood wielding his makeshift spade like a cashcrom (the traditional peat-digging spade).

I've also realised that in my passion for digging there probably lies a subliminal connection with the past behind us and the earth beneath us, another two of my passions in life. I don't just mean that by digging archaeologically we can connect with the remains of past societies, but that through the act of digging itself we can experientially connect with past peoples. From prehistoric hill forts to Roman roads, fenland drainage, ditches, tunnels, dykes, cellars, causeways, canals, midden pits, foundations, irrigation systems, railway embankments, moats, road sidings and mines, it's clear that *Homo sapiens* has been as much a digger as a maker. In fact, almost every period in our island's history is characterised by some form of digging.

Since retiring from full-time archaeological excavation I've found solace, and a renewed passion, in a particular type of digging, and one from which I've learned an enormous amount about myself and our human past. It is the digging we do for food. Few people have cleared virgin ground by hand in order to grow food. I have, and as far as my lower back was concerned, four times was three times too many. Digging over a vegetable patch is hard enough but such beds have invariably already been broken, and the soil in a well-worked garden plot, if it had reasonably high levels of seed germination, will already be fairly stone free. A good grower will also have kept the levels of organic fertilising matter high, making the soil soft, spongy and responsive to the turning spade.

Virgin ground is a different prospect. In all my years of historical farming and petrol-free gardening, I've found that there is one tool, the mattock, that ranks above all others when it comes to turning a wilderness into an area that can produce food. Other garden implements flatter to deceive. The spade, for all its sharp and neat lines, is redundant when even the smallest stone, cushioned by compacted subsoil, renders its slicing motion ineffectual. The fork, while successfully navigating around such stones to reach the required depth, often lacks the strength in the tines to lever the earth free. We've all seen the twisted, buck-toothed tines of garden forks that have been asked to do jobs for which they are little match. But drawing the mattock from the toolshed and slinging it over your shoulder as you march off in the direction of the ground you intend to break represents not only a commitment but also a recognition of the work entailed. Few undertakings in the world of manual labour, perhaps with the exception of ditch digging and quarrying stone by hand, place man closer to the base works of humankind's evolutionary journey. If you really want to get close to the past, as close as you can possibly get, then take a patch of unforgiving land and attempt to feed yourself from it. Doing so opens a window into the eternal struggle of human existence.

For my own part, I was foolish enough to refuse the loan of a petrol-powered rotavator, wilfully blind to the time it would have saved. Time is money. Yet, this was a journey I was determined to make. Breaking ground, I felt, brought me viscerally into direct contact with the past. I wanted mud on my boots. But not because I'd traipsed out on a jolly ramble to survey the archaeological remnants of some prehistoric fields in the landscape. What I wanted was to dedicate the time and effort to recreating my own version of them.

So what is a mattock? Dating back to the Bronze Age, a mattock resembles a pickaxe but with wider blades of similar size set in opposing planes at either end of its head. On one side a vertically set blade acts as a kind of axe while the horizontal blade opposite takes more the form of an adze. Usually, it's the horizontally set blade that sees the bulk of the work, but every now and again a disruptive root submerged beneath the path of your work requires severance. Here's where the vertically set blade comes into its own. Mattocking the ground is a relentless process. Working in three-foot strips, you gradually plod your way up and down the plot. Each clod broken free of the ground is the result of lifting a seven-pound block of iron above your head and bringing it crashing down, shattering the earth beneath your feet.

It's not long before your hands are on fire with blisters. The sweat stings into the creases around your eyes and a numb, menacing twinge develops in your lower back. This is a job that tames you. Having started out with all the vigour of youth, boldly hammering away at the ground, you very quickly tire. The swinging motion becomes wilder and less controlled as your muscles weaken. If you're not careful, the mattock will drop short of your target, skid off the surface and swing dangerously close to your shins. You stop. Panting, you survey the pitiful results of your power burst. You pace around it, breathing heavily and mentally calculating how much energy you've expended against the small patch of ground you've covered. Choosing not to dwell too long on that, you then start again. Gradually your pace slows and, like a horse brought in from the plains, you are tamed into the work. You resign yourself to it. Your breathing moderates as you become methodical, more controlled. This is a marathon, not a sprint. Your breaks are regular, but short. You give yourself enough time to straighten up, stretch your back and clean the blades of the mattock with your raw hands.

I did a lot of mattocking while working as an archaeologist. Archaeology in the field, at the actual point of excavation, has strange parallels with basic agricultural digging. In most modern scenarios a machine would be used to dig out, say, a six by six-foot-square pit. Indeed, most construction companies these days are so scared of litigation that they won't countenance any other method for fear of accident and injury insurance claims. But if that six by six-foot-square pit just so happens to be a medieval midden pit, packed with precious archaeological data, then it can only be excavated by hand. And because of the necessary pressures placed on archaeological processes by the construction industry – for whom the archaeology is usually undertaken in the first place – it's agreed between construction engineer and archaeological supervisor that a happy medium, somewhere between the hand trowel and the mini-digger, can be employed: the mattock.

However, swinging a mattock in the service of archaeology is rather different from using it to break ground. Because care and attention are required when excavating valuable archaeological deposits, and because you're often working in confined spaces with other archaeologists, the business end of the mattock is rarely lifted above the head and swung wildly down. The technique becomes more one of a chipping away, of retaining enough control to pull out of the hacking motion should you expose what might be an important archaeological find. Even so, it's just as laborious, and more so if you have to spend all day bent double. Luckily, when breaking virgin ground such caution is not required and you can use the weight of the mattock head to your advantage. My father's mantra – always 'let the tool do the work' – rings in my ears whenever I wield a mattock. Controlling the speed and direction of the swing, guiding the seven-pound iron lump down with a touch of added force is invariably enough.

During this period as a jobbing archaeologist I found myself on one particularly challenging site. While we were under pressure to get the job done so that the building of a vast commercial complex could get underway, an old Irish construction worker gave me a sage piece of advice. Seamus was part of a team putting in service and foundation trenches all around us, a measure of how pressured the situation was that this was happening before all the archaeological investigation had been completed. He'd been watching me excavate a series of vast medieval pits on the south bank of the Thames opposite the City of London. While he sat comfortably in the air-conditioned cab of his mini-digger, there I was in the baking sun hacking away with my mattock, feverishly trying to get the job done on time. Seamus was no stranger to this type of work. Over an after-work pint he told me about his youth and his emigration to England in search of employment in the late 1960s. He'd found work in the highways and construction industry, as many of his fellow Irishmen had at that time, and had learned the trade the hard way, in an age before mechanical diggers. As a consequence, he'd probably forgotten more than I'll ever know about digging holes by hand. His advice was simple and can be extended to so many aspects of life.

'Are you right-handed, Alex?' he asked.

'Yes, Seamus.'

'Do yous want to end up like the hunchback of effing Notre Dame?'

'No, Seamus.'

'Well, for the Lord's sake, swap the effing mattock over to your left hand every ten minutes, lad. With every blow you're twisting your effing spine.'

I paused for a moment, considered, and duly gave his advice a try. At first it felt clumsy, and I wasn't getting nearly as much work done. I tried to keep it up but gradually the lesson faded and I resorted to my old ways. About a year later, however, Seamus's advice came back to haunt me. During the excavation of a vast Iron Age ditch section on a site on the outskirts of Worthing in Sussex, I twisted my upper body and spent the next three months in agonising pain. Now I religiously swap hands at regular intervals with most tools, whether sawing wood, raking leaves or chiselling timber. I even clean the teeth on the left side of my mouth with my right hand and the teeth on the right with my left hand. All because of Seamus.

So it was with this expert advice that I set out with my mattock and a substantial bottle of water to the virgin land I had decided to grow food on. Mattocking is far from the end of the process. I only realise this now, five years on, when the very patch I laboured over is getting close to yielding vegetable crops similar to an adjacent patch of ground that has been under the spade for decades. Once the ground is broken, then comes the challenge of working it down. In today's agriculture, this is achieved through numerous phases of cultivating and harrowing, but on my plot (or my hypothetical prehistoric field system) these processes were substituted with digging and raking.

The mattock and the spade take the role of the cultivator and plough and serve to break up the ground into clods. The rake carries out the work of the harrow, drawing the unearthed root matter and stones from the topsoil. In the fields, different types of harrow are used to work down clods, disentangle root matter and collect up stones. For instance, a spike harrow (a series of downward pointing spikes set on a frame) might be used in the first few passes, and as the soil is gradually reduced to smaller particles, with the larger weeds and stones being gathered up, chain harrows (a smoother frame of loosely bound metal links) are used to create a more even tilth. Similarly, in the garden plot a fork might be used for clod-breaking, while various scales of rake can be used to draw weeds out and pick off the larger stones.

On land that has already been worked down such processes might be undertaken once every three years or so to prevent compaction. But the breaking of virgin ground requires the repeated forking out of all root matter, breaking up of clods and removal of any stones for at least the first few years of its use. This is because no matter how thorough you are in your first attempt, some always slip through the harrow, and it's also good to let the ground-breaking processes of frost and thaw help in the work. As a result of repeatedly conducting these operations on my plot, a substantial pile of stones and weed roots had developed at the top end. Bonded together by a matrix of earth, the by-product of my endeavours was now a mound. One morning, as I considered whether I had the energy to do yet another round of raking and hoeing, the prominence of this edifice caused me to muse on the origins of one of the most iconic monuments of the British prehistoric landscape: the barrow.

In modern archaeology, barrow is a term used to refer to a range of landscape monuments of varying sizes and scales but all consisting broadly of mounds of earth and stone. In a bid to define a typology, archaeologists have applied a plethora of descriptive terms to barrows, and some of my favourite come from an early attempt at characterisation in the nineteenth century by the eminent antiquarian Sir Richard Colt Hoare. There are long barrows, bowl barrows, disc barrows and bell barrows – to name just a few – and while most are completely surrounded by ditches, others have causeways crossing these ditches in a clear indication that symbolic passage between the centre of the monument and the outside world was considered important.

Many barrows were erected as funerary monuments and ritual sites for a range of mortuary practices, including the interment of cremations. The people honoured in this way were obviously of high status and this is occasionally reflected in the variety of luxury and prestigious goods they were buried with in a central chamber. Some excavated examples of barrows have been found to be without central chambers or evidence for burial – perhaps because the function was not fulfilled, or because the mound was providing some other role in the landscape.

Broadly speaking, archaeologists have applied a chronology.

Long barrows appear to represent the earliest British examples, dating to the Neolithic era. Around three hundred survive, some in better condition than others, and the fine preservation of examples at West Kennet in Wiltshire and Wayland's Smithy in Oxfordshire serve best to illustrate the size and function of these monuments. Then, in the Bronze Age, smaller circular barrows became the norm and are far more numerous and widespread throughout the British Isles. In some places they can be seen grouped together to form barrow cemeteries, and more often than not are associated with earlier monuments such as henges. At Stonehenge, for example, 260 such barrows can be found within a radius of just under two miles. Towards the end of the prehistoric period at the dawn of the first millennium AD, this form of monument and burial practice appears to die out, but the story of the barrow doesn't end there. As monuments whose presences linger as vestiges of a past age, barrows are often repeatedly reused and adapted.

The early Anglo-Saxons, fifth- and sixth-century invaders and migrants from northern Europe, often reused Bronze Age barrows for the burial of their own rulers in a desire to associate themselves with this embedded ancestral landscape (and perhaps reflecting an insecurity over their own right of tenure). But the Anglo-Saxons also introduced a tradition that had survived in their Germanic heartlands of constructing barrows for ceremonial and burial purposes. For them, the barrow remained alive and well as a symbolic edifice of power and status, and they commonly associated pagan deities with their own new-build barrows and those of the British prehistoric landscape. Ritualising the landscape in this way imbued it with meaning. To my mind, no place makes this plainer than the naming of the long barrow in north Wiltshire as Woden's Barrow. There were also functional uses in the early medieval period, and it's apparent that they operated as convenient markers against which to draw up property boundaries and as prominent (and meaningful) places at which to muster for the sake of assembly, governance and justice.

In the later Anglo-Saxon period of the tenth and eleventh centuries, as Christianity tightened its grip on the establishment and spread its tentacles into the recesses of rural life, such monuments needed to be placed more firmly within a stricter ideological framework. If they were to remain revered then they needed to be adorned with motifs of the new religion, to become the focus for conversion processes and marked with churches and consecrated ground. A classic example can be seen at the church of St Michael and All Angels in Berwick, Sussex, where a large Bronze Age barrow squats stubbornly in the graveyard – an immoveable relic of a past belief system.

But for the barrows beyond the immediate reach of the church, still enthralling the local people with their pagan associations, an altogether more sinister transformation might be required. So it is that associations with the Devil become common in the medieval period, and again from Sussex, the Devil's Humps are one of the most vibrant illustrations of this. This deviant character is not only expressed through nomenclature, however, as a selected few barrows became places of summary execution and the burial grounds for society's outcasts. If you were unfortunate enough to find yourself sentenced to execution in the first place, in the minds of the medieval jury your repentant pain and suffering were only just beginning. Burial in association with such a demonic monument was a declaration that you would spend an afterlife consigned to the eternal tortures of hell.

It is perhaps as a consequence of this demonic legacy that barrows slipped into the very margins of society in the Early Modern period, retaining only a lingering, folkloric and almost comical association for the Tudors and Stuarts. However, with a more scientific understanding of the world emerging in the eighteenth and nineteenth centuries, barrows once again attracted attention. Antiquarians like John Aubrey in the seventeenth century and William Stukeley in the eighteenth were quick to place barrows within the grander narratives of humankind's origins, while the likes of Richard Colt Hoare and General Augustus Pitt Rivers, at either end of the nineteenth century, did much as the forefathers of archaeology to apply modern archaeological techniques in their pseudoscientific excavations.

Archaeology, history, place names and folklore have all helped tell the colourful story of the barrow and yet, as is often the case, it's the origins of these monuments – the moment of inception – that elude us. Why did this form of monumentalism emerge in the landscape of Britain in the fourth and third millennia BC? Entering the minds, exploring the thought processes and imaginations of people in an age before historical texts revealed the past's own self-awareness has always been the greatest challenge facing archaeologists of the prehistoric period. We can describe, scientifically, the physicality of the past but lending meaning and human understanding to those physical descriptions necessitates an interpretative leap.

It was in the process of working down my first patch of ground and subsequent years of clearing it of stones and weeds that I made my own interpretative leap. I've been lucky enough to participate in the excavation of three prehistoric barrows, all in the county of Wiltshire. The first was of a barrow that had already been clumsily exhumed in the nineteenth century, revealing the skeletal remains and grave goods of a Saxon princess; the second was incidental to a late Romano-British farmstead we were excavating at the time; and the third was to assess the level of damage being done to such precious monuments by an unchecked badger population. In each case, it was clear that the fabric of these barrows was a result of numerous phases of construction rather than of a single Herculean effort.

They had been revisited over the course of generations, with each generation adding layers as part of their ritual interrelationship with the mound. But what was at the heart of that process, and what meaning could be derived from these repeated acts? Revisiting a mound that was already the site of burial may have acted like a form of currency, a kind of prestige bank, where the size and scale of the mound was in effect a reflection of long-term remembrance of a significant ancestry. For those who had little in the way of material wealth to offer up in reverence, devotion could be communicated by the amount of earth one was prepared to move.

I spent about three years working my own small field. In my successive phases of digging and raking, I created a large pile of stones, loose earth and weed matter. With my every working down of the land and subsequent removal of debris another layer was added, and the soil of my land improved, increasing my yields. As I undertook each episode of clearing I vowed at some point to remove this stony mass to another part of the garden that required infilling. Yet on a winter's day, when all the surrounding vegetation in my garden had died back, this mound took on a visual prominence. What's more, it had become something of a symbol on which to tag my labours. In it was embedded all the back-breaking toil, the hardship and, yes, the small heroism of creating a patch of land that would now produce food for my family, but also for those who followed after. The hard work had been done, for the ground had been broken, the stones removed and the person who had done so was also the father of this barrow. Both field and barrow were now my legacy. To the outsider, it was just an unsightly and obstructive pile of earth in what I hoped was an immaculately tended vegetable patch. But, in my own private world, it was a symbol of prestige.

It spoke loudly to me that this was what you had to do if you wanted to take unbroken ground and work it down to produce food. It was a badge of honour.

Might barrows have first been the product of such work? Might time, and the recognition of what it took to break ground, then have imbued them with a similar sense of achievement and even prestige? It's perhaps no coincidence that their appearance in the prehistoric landscape is broadly contemporary with the wider-scale adoption of agriculture and the creation of our earliest field systems. Barrows as monuments were undoubtedly achievements but, critically, so too were the fields that surrounded them. On a grander scale, in a hierarchical society, it might be easy to envisage how someone who had the power and authority to commission a large-scale ground-breaking project might be quick to commandeer the products of that labour – not only the harvests yielded by the fields but also the mounds that resulted from land clearance.

Of course, I'm not suggesting that all barrows are the results of agricultural clearance. But I am exploring the notion that the beautifully crafted and ritualised constructions of the late Bronze Age had ultimately emerged from an earlier tradition of clearing ground. Clearance cairns are a known phenomena in the British landscape, particularly in stone-rich landscapes. But because they lack associated artefacts, such as grave goods or skeletal remains, they are often difficult to date scientifically. It's also very much the case that in the various phases of agricultural improvement in the last three centuries, clearance cairns have necessarily been produced.

Perhaps we should cast a more critical eye on those clearance cairns of a likely prehistoric date, and see these monuments as part of a transitional stage, as stepping stones on the path to the deeply ritualised and stylised barrows of the later Bronze Age. They are the keystones, the edifices that allow us to go from an untamed landscape, the stomping ground of the gatherer-hunter, to the cleared ground of the first farmers. In this context, clearance cairn comes across as far too pragmatic a term for such important constructions. It feels as though it robs them of their symbolic association and the social significance they gained through their role in a momentous phase of landscape development. It almost belittles them.

The barrow and the field march together across the evolving landscape of the Bronze Age. We might not see one as the product of the other, but the barrow helped, through reinforcing the ideological mores of the day, to create the social cohesion that underpinned the production and maintenance of field systems. And as increasing numbers of trees were cleared to create agricultural space, so the visual spectacle of the sweeping landscape came into being. What the barrow and the field represent are one and the same: humankind's newfound ability to manipulate the material world to our own advantage. In the land's domestication, man was manipulating the earth in a way he had never done before – working with materials to improve and construct a terrain that suited him. This new world replaced the graceful art of the hunt and the pastoral motions of the gatherer with the brute force of the ground-breaker and the farmer. But in this relentless taming of the wild, humankind was itself domesticated, tamed by the craft of digging.

14

BASKETS AND BOATS

MY FAVOURITE BASKET is a small picker's basket I purchased in France a few years ago in circumstances which, looking back now, I consider to be quite poignant. In Britain we have car-boot sales, in the US they call them yard sales, but in France the vending of second-hand goods and unwanted junk usually takes place at what is called a *vide grenier* (with a literal meaning of 'empty the attic'). For an Englishman starved of opportunities to buy vintage farming kit in his home country such events were unmissable. France retains a much more tangible connection to its rural past, and over the years I've managed to pick up a marvellous collection of tools, utensils and memorabilia, the highlights of which include hand-forged billhooks, various old farm tools, a shave horse, a 1930s copper-bodied sprayer, a gorgeous penknife that I still use today, a pair of late nineteenth-century pince-nez spectacles, a complete run of *Paris Match* magazines from January 1952 to August 1962 (to use as wallpaper), and an apple 'scratter' box (for pulping apples before pressing). I could go on. I think the only person I know who loves a *vide grenier* more than me is my father, who has a radar-like awareness of their frequency and locations for the whole Sarthe region of France. So when Dad mentioned popping in to one en route to lunch in Le Mans one day, I – and the rest of the family – jumped at the chance. Nothing whets the appetite for a three-course *plat du jour* than a jolly good rummage through boxes of someone else's junk.

But France was a changing country and the *vide grenier* at the village of St Saturnin was a reflection of that. As soon as we arrived it was obvious that a morning poring over a broad selection of antique farm tools was not going to happen. The new-build houses and the young professional demographic were a clear indication that this was suburbia, the urban sprawl of Le Mans as it crept out from its medieval core subsuming its satellite villages. The inhabitants were city types, aspiring metropolitans who had long ago turned their back on the old ways. And their material cast-offs, spread out on tables and lawns in front of their houses, revealed that. Toys, buggies, outgrown kids' clothes, outdated electrical equipment, DVDs, CDs, videos, budget kitchenware – that which was once marketed as new and fashionable now looking cheap and insubstantial. I barely broke stride as I circulated round the stalls desperately scanning for anything of value.

It was only at the very last stall that I glimpsed something of note. I moved in closer to check that my eyes weren't deceiving me and was delighted to confirm three beautifully made baskets stacked into each other. A woman, perhaps in her late fifties, stood marshalling the stall while a very old man sat back from the table, leaning on his walking stick, gazing blankly into the middle distance. I gestured for permission to pick one up and was enthusiastically encouraged. As I turned it over in my hands, admiring the form and the standard of craftsmanship, the woman said it was the work of her father (the seated elderly man), and that when he had the energy he spent his evenings making these baskets in the way he'd been trained by his grandfather. I realised very quickly that I'd struck gold. Here was a basket of some pedigree. Although it was brand new and unused, the skill set, the craft and the intangible heritage that lay behind its making were of the finest calibre. At only thirty euros it was an absolute steal and I snapped it up immediately to add to my collection. As I wandered back to the car, relieved that my morning hadn't been a complete waste of time, I was conscious of the telling contrast between the plasticised junk belching out onto the streets of St Saturnin and the beautifully handcrafted object tucked under my arm.

URING THE COURSE of the past six years, I've spent
some time, as a patron, representing the Heritage Crafts
Association. I frequently canvass support for the group in the
hope of inviting financial backing for traditional craft skills. On
more than one occasion, when raising the need for more funding
to support skills that are genuinely dying out, I've been met with
a derisive smile and a prejudiced response, along the lines of:
'Who would want to fund a bunch of bearded, hippy basket
makers?' The insinuation is clear: crafts are hobbies, of which
the most symbolic is basketry. To such dismissive remarks, I
smile back and launch into some of the statistics of the heritage
crafts industry: over 80,000 firms in the UK today employ more
than 200,000 people, with the sector having an annual turnover
of nearly £11 billion, contributing a whopping £4.4 billion in
gross value to the economy. While this tends to shut up the
economically minded sceptics, that basketry is singled out as
the skill that typifies the perceived non-economic viability of
traditional making always leaves me with a sour taste in the
mouth. This is because basketry not only represents one of
the most ancient of craft technologies but it is arguably one of
the most long-lived. It may not be the best moneymaker but it
deserves respect.

Exactly how old basket making is might never be resolved
owing to the organic nature of the material. Only in exceptional
conditions would artefactual evidence survive. But when
baskets show up from the Mesolithic period (broadly 10,000 BC
to 5,000 BC), in rock paintings of central India, as waterlogged
wicker fish traps in the tidal zones of European rivers, and as

desiccated remains from Egyptian desert sites, they are already highly sophisticated and fulfil a remarkable range of functions. The craft certainly pre-dated pottery – we know this because some Neolithic pots have stylistic cord impressions, an effect that imitates the characteristic weave of a basket. This may be pure skeuomorphism, a concept that sees new materials, in this case clay, used to replicate and resemble organic antecedents. But it could also hint at the practical means by which pots were produced in the first place: baskets were the outer retainer within which the pots were supported as they were formed and built up by hand. So they are ancient – very ancient – and very likely one of the first true human crafts.

What is truly remarkable, however, is that humankind still makes baskets today, to fulfil important economic, social and cultural functions the world over. For the developing world, baskets symbolise the resilience of indigenous culture – the ability to make beautiful practical receptacles from locally sourced raw materials – and in the developed world they continued to play a fundamental part in everyday life well into the twentieth century, despite industrialisation. This is because of the complexity of the craft itself and the inability of designers and engineers to develop a machine that is more effective and more cost-efficient than the human being. It's true that cardboard and plastic have taken over as materials to make cheap, disposable packaging, and that you can purchase an off-the-shelf basket-making machine from China that casts and moulds plastic crates. Laundry baskets can now be made from plastic, but there is no weft and no warp; these are not *cræfted* baskets. True basket weaving was, and remains, the original 3D printing. For any size, any shape, any function, there was usually a basket for the job.

In a survey of the rural industries of England and Wales undertaken on behalf of the Agricultural Economics Research Institute in 1926, there were estimated to be over two hundred varieties of baskets in production. Add to this the hobbyist, the part-timer and the cottage basket maker, and it's likely that at the dawn of the modern world there were more types of baskets being made in Britain than at any other period in our island's history. You would have had your regular 'pickers' or 'shoppers', baskets carried on the arm and ranging in size and shape according to what is being picked or carried. Flower baskets, for example, tend to be flatter and open at either end to accommodate long stems. Shoppers, covered with lids, become picnic baskets. Squared off and with side handles, they become hampers. Threaded with leather straps they can be mounted on bike or pony as a pannier. Table baskets can be used to display fruits, vegetables and bread. The sleeping dog has a basket to keep it off the cold ground. The racing pigeon has a basket for carriage. The broody canary has a basket to nest in. Laundry baskets can be containers for collecting dirty linens or open baskets for carrying from washing drum to line. A log basket holds the fuel for the stove; a peat-carrier brings it in from the store. The potato basket must be strong and robust; the waste-paper basket can be light and flimsy. The lobster pot needs to cope with submersion but must be light enough to be pulled from the deep; the egg basket curves inwards on its rim to stop the precious cargo from rolling out. And then there are the furniture items: baby's cribs and rattan, raffia and wicker chairs. This cursory list doesn't do justice to the full range of functions that basket making fulfilled at the turn of the century, but namechecking a handful does at least demonstrate that we couldn't have survived without the weft and warp of the basket-maker's craft.

Well into the twentieth century the ubiquity of the basket was all down to the fact that for lightness and strength it was unsurpassable. This point was most emphatically brought home to me during a visit to the Fleet Air Arm Museum in Somerset. As I wandered around the galleries, taking in the various aircraft that had defended Britain's interests from the perilous starting position of a ship's deck, I naturally gravitated towards the earliest planes in the collection. In the first decade of the twentieth century, the field of aviation was very definitely the most innovative and most advanced technology of the day, but when the Short S.27, which in 1912 was the first aeroplane to make a successful take-off from a moving ship, needed a lightweight but robust seat, for want of a better material the designers turned to basket weaving to construct a squat wicker chair. Examining up close the rather fine canework seat, it seemed to me the most literal instance of 'warcraft'.

IN BRITAIN, THE most renowned type of basketry – commonly termed wickerwork – is that which interweaves fine rods sourced from the willow family. For many reasons willow is the preferred material for the basket maker. Although a hardwood it's extremely flexible and lightweight, properties that are best demonstrated in its use for the making of cricket bats. But willow is also an inveterate grower in wetland environments. Cut a young willow tree back to its roots in winter and several vigorous straight shoots will have taken its place the following spring. Willow also propagates incredibly well. You can take a small section of rod, perhaps six to eight inches long, drive it into the ground in late autumn and, as long as the earth remains moist throughout the winter months, a young shoot will emerge in the coming year. Given time to grow and establish itself, this root will go on to provide willow rods for at least a generation.

It can't have taken long for humankind to observe all these qualities and take steps to harness the power of willow. I feel sure that as early as the Mesolithic, societies would have propagated and managed wetland environments to stimulate and encourage willow as a resource. The proliferation of references to 'withy' (the Old English word for willow) in some of the earliest historical sources for the English landscape suggests that by the tenth century they were suitably pervasive to be worthy of mention. Over time, in the reciprocal relationship between landscape and craft, as willow has been used to shape our baskets, our need for baskets has caused the shape of the landscape to change in order to produce the raw materials. This is most vividly demonstrated in the landscape known today as the Somerset Levels. The Avalon of legend, the last resting place of King Arthur – the marshlands along the course of the River Parrett and its confluence with the Rivers Isle and Yeo – remains today a watery, otherworldly and magical place. This is the home of Glastonbury, with its abbey, tor and festival, and the location of Athelney, the place Alfred the Great fled to at his lowest ebb, to seek sanctuary from the invading Vikings before mustering his forces to defeat them.

Drive along the M5 motorway today and it doesn't feel very Arthurian. You'd scarcely register the low-lying fields and the network of dykes, ditches and canals – or 'rhines' as they're known in the local dialect. But the floods of 2013 were a timely reminder of how low-lying this land is, and how well drained and managed it had been in past centuries. In fact, one of the suspected reasons the floods proved so catastrophic was that as the industries of the wetlands – the seasonal grazing of sheep and cattle and the willow and water-reed harvest – have fallen into decline, so too has the management of the man-made watercourses that sustained them. The monks of Glastonbury Abbey were the first recorded constructors of such waterways, but it seems likely that the Romans had a hand in some of the earliest engineering works. And this suitability to grow willow made the Somerset Levels a focus for willow production well into the twentieth century.

This is a land that has the capacity to take 16,000 propagated plants per acre to establish withy beds that are productive for a period of forty to fifty years, although it's not unheard of for some beds to be in service for over ninety years. Strong varieties like the Champion Rod and the Black Maul became the favoured willows in an industry that was to have a profound impact on the local economy. By the seventeenth century, demand for good willow cane was sufficient enough that some landowners realised the economic benefits of growing it intensively. Book Five of John Norden's *Surveyor's Dialogues* of 1607 imagines a conversation between a surveyor and a bailiff, a literary device used to convey best practice, and the bailiff remarks how his willow beds yield, 'greater benefit yearly, acre for acre, than the best wheat'.

Yet willow is far from the only material used for making baskets in Britain, let alone the rest of the world. Traditions grow out of landscapes. Certain geologies produce certain soils, which in turn produce certain vegetation, which when harvested produces certain raw materials that are best suited to certain baskets. On drier soils, for example, where willow is in shorter supply, hazelwood makes for an ideal replacement. It would be madness – and distinctly un*cræft*like – to import bundles of willow over long distances by road and rail when suitable materials can be sourced from one's native hedgerows. So, on a local and regional basis, any material would be considered, providing it was suitably fibrous and retained its strength once in desiccated and processed form. Even my arch-enemy, the dock-leaf plant, has been found to be of use in Scotland, where the dried stems of this most vigorous of weeds were used to make baskets known as 'dockens' or 'kishies'. It is the social and economic logic of matching a locally abundant material to a local need. We have already seen how a basket made from twisted straw (a by-product of cereal harvests) bound with bramble cane (an abundant hedgerow species) made an ideal beehive, due in part to its properties of insulation.

But of all the materials I've seen used to make baskets, the one that has astounded me most was oak. The English oak conjures up visions of huge majestic trees with giant trunks and boughs felled to be used for ship and house timbers. Only the other day I was clambering through the voids above the nave of Salisbury Cathedral and admiring the timbers that supported the pitched lead roof above. The tie beams stretching from one wall to the other across the entire width of the nave were a massive fifty feet long and getting on for two feet thick. How on earth could such a material be used to make a basket? The answer lies in the spelk or spale basket. Known in other parts of the country as a whisket, swill, slop and skip-basket, the sourcing of the raw material is the key to its production.

Spales are made from oak that is rent (split along its grain). But it's not the large construction timbers that are treated this way, it's the poles harvested from a coppiced or pollarded oak. Oak behaves like any other tree when chopped at its base (coppiced) or at the head of its trunk (pollarded): new shoots will emerge and, given a good ten years to grow, will develop into fine poles. These are then bark-stripped and split down their length into quarters using what's called a 'froe' or a 'dole axe', a name derived from the Old English word *dál*, meaning to share or dole out. These thinner lengths, still nearly two inches thick, are then boiled down for an hour or so to open up the grain. This is when things get clever, and the point at which I have to mention a certain master spale-basket maker, a name that is never far from the lips in any discussion among the green woodworking community on best practice.

About seven years ago I had the privilege of spending an afternoon in the company of Owen Jones as he demonstrated how to make a spale basket from scratch. It was his method for producing 'spelks' (thin strips of oak) that was most astounding.

Taken directly from the boiling bath, still piping hot, he would make an incision at the top of the length and then use his fingers to peel away strip after strip along the grain of the wood. When the extreme temperature became too much, bracing the length between his legs, he would slap his hands against his knees to knock out the heat, before returning to the splitting. It was astonishing to watch, and at the end of what can only have been an hour or so, he had a substantial pile of two-millimetre strips of oak lying at his feet. Separately, a single piece of hazel would be moulded around a former to make a rim or 'bool', the frame around which the spelks, kept moist for pliability, would be woven. The stouter ones would be used for the warp and the smaller, known as 'chissies', for the weft of a basket that was remarkably strong and lightweight. Owen was most proud of the fact that should one of the spelks become fractured they were easy to replace, which meant that these baskets, kept in the right conditions, would have a considerable working lifespan.

I loved working with this basket. There was an ergonomic intelligence to it in terms of its size, the location of its handles, its weight and its fittingness to do the job required – which, in my case, was to carry sliced root vegetables from the root-slicer to the pig sty on a twice-daily basis. As it became my go-to carrying receptacle I learned to trust and value it, and so began to associate its aesthetics with good design. What I marvelled at most, though, was the level of structural integrity that was derived from the manner in which it had been made. You can now take postgraduate level degree qualifications in structural integrity where component materials are subjected to exhaustive methods of scientific observation and testing. Owen Jones relied on tradition, a lifelong apprenticeship and a *feel* to achieve his results. His was a palpable connection with the tacit knowledge associated with one of the best building materials known to man – oak. I realised, however, that this form of construction

– and hence the levels of structural integrity – wasn't unique to baskets and that it could be applied to a range of building uses. Even houses and boats could be made using the techniques of basketry. Can you imagine living in or indeed crossing the ocean in a giant basket?

In the mid-1980s, deciding to undertake improvements to his land, a farmer in Glenarm, County Antrim, took to bulldozing flat a large mound that sat incongruously in an area that was otherwise well suited to being ploughed-up for cropping. He soon realised that on top of this mound lay a *ráth* – the Irish term for a ring fort – and subsequent archaeological excavations exposed one of the most significant historic Irish sites of modern times. The recovery of glass, leather, metal, ceramic and textile crafts suggested a reasonably high-status site, but it was the evidence for the construction of the settlement structures that was so remarkable. The walls comprised wicker weavings – but not in the way that a wattle and daub wall infills between larger timber frames. The whole structure was built of wattlework – hazel rods woven through upright stakes to create a continuous circular wall. What made this dwelling so effective was that a second wattle wall created a cavity within which moss, heather, straw and other organic materials were compacted to aid insulation. The structures – some of which were nearly twenty feet in diameter – survived only to a height of twelve to fifteen inches, making any reconstruction of their full superstructure hypothetical. They could have been domed or worked up into a cone. Either way, their inherent strength – their structural integrity – relied, like the spale basket, on the interweaving of relatively small and thin rods of wood to create a composite structure of considerable strength.

The choice of construction may well have been down to social and cultural conditions: this is what you do when tall trees of oak, beech, elm or ash hardwood from which to produce large structural timbers either don't exist locally or you don't have permission to fell them. It could also have been about access to certain technologies: you build in this style when you can't get your hands on bespoke carpentry tools such as forged chisels, adzes and augers. But it might also have been a construction style born out of choice, tradition and tried and tested methods. Clearly, these structures worked and were popular. Thirty circular dwellings were uncovered, occupying a period of over three hundred years (*c.*650–950 AD), within which it's estimated that at any one time three or four were in use. Snug, cosy and incredibly resourceful – living in a basket couldn't have been that bad.

But what about crossing an ocean in a basket? That can't have been much fun, but it happened. The great tradition of *peregrinatio*, leaving the security of one's homeland to wander in the service of God, was popular among Irish monks in the early medieval period. They were driven by their desire for self-discipline and austerity, and considered there to be no greater hardship than casting oneself off into the ocean and placing one's life in the hands of the Lord. For such a journey, the vessel of choice was a coracle. In *The Voyage of Snedgus and Mac Riagla*, one of the three ancient Irish *Immrama* ('voyage') tales, the two heroes abort their mission to return to the holy island of Iona and, 'as they were in their coracle, they bethought them of wending with their own consent into the outer ocean on a pilgrimage'. Foolhardy perhaps, but the truest form of ascetic renunciation for the religiously inclined of seventh-century Ireland. It's only when you realise exactly what a coracle is that the folly of these monks takes on an extra dimension.

While there are archaeological examples of coracle-type vessels from as early as the Bronze Age, it's not really until the twelfth century that historical sources confirm explicitly the size and composition of a coracle. Gerald of Wales, in his *Descriptio*

Cambriae of 1194, gives the most comprehensive description: it's made out of withies, has a rounded behind, with a pointed front that forms a triangle shape. Some indication as to its size is given when he writes of the 'primitive habit' of the Welsh, 'of carrying their coracles on their backs' to and from the river where they fish. At around one foot deep, six feet long and four feet across and weighing under twenty pounds, I would consider any vessel that can be used effectively to fish, *and* be carried about one's person, to be a result of sophisticated rather than primitive design. But Gerald also indicates how precarious the coracle could be in water when he tells us that a salmon, by landing in the vessel and thrashing its tail, is enough to topple it over. For the waterproof outer lining, he informs us that untanned animal skins were used. This waterproofing is detailed in another Irish early medieval source, *The Voyage of St Brendan the Abbot*, where we learn of sailors taking fat or butter with them to oil and protect the skins.

So, for the Irish monk intent on the near suicidal endeavour of *peregrinatio*, the coracle was the perfect vessel. Low-status, humble and thus pious, it was also without a keel, which on most small vessels was the only means to mitigate the lateral movement of wind and current. Even if you wanted to steer a coracle in the open seas, you didn't have much choice in the direction you were taken. It was up to God. But it was this design feature, as well as its lightness and simplicity of construction, that saw Irish and Welsh fisherman using coracles – or *curraghs* as they're known in Ireland – in the fast-flowing rivers of their homelands right up to the twentieth century. In rivers like the Corrib in Ireland or the Teifi in Wales, the currents are fast and furious and lightness of vessel is imperative to skim over the rapids from one bank to the other without being washed downstream. Despite Gerald of Wales's accusations of primitiveness, the coracle is an ingenious contraption that could be made with limited resources and used to secure a critical part of the human diet. It is as much a *cræft* as it is a craft.

I could explain in detail the making of a coracle, but through describing how Owen Jones made a spale basket I already have. Interweaving split or rent timber is known from surviving examples, particularly when reinforcing the gunwales. But the withies – willow rods – mentioned by Gerald of Wales continued to be as popular in the twentieth century as they were in the twelfth. In later times, animal hide was replaced with coverings of canvas, waterproofed with tar and pitch, but the essential principles of coracle construction remained the same – for centuries.

IN THE DEVELOPMENT of the coracle, it may well be that some bright spark back in the Bronze Age, having lined his basket with oiled leather, realised that anything that held water in could also keep it out. All he had to do was make a basket large enough to hold him. It is hypothetical, this simple evolutionary model, but interesting nonetheless that the standard size for a coracle might have been determined by the size of a cow – the maximum size of vessel that could be made from one cow's hide. For another type of basket, these evolutionary steps seem to have passed in the opposite direction: a specific technique for creating a boat is scaled down from one of the finest craft moments in human history to the making of a simple and traditional basket of a very local variety. This is the story of how certain aspects in the construction of a Viking longship can be seen fossilised in the form of Sussex's famous basket: the trug. Add an 'h' to the end of the word trug, and an 'o' before the 'u' and you have pretty much the same word in its modern form. But as well as 'trough', the

Old English *trog* also appears to mean 'tub', 'basin' and 'boat'.

It wasn't until after I'd left Sussex in 1995 that I really started to connect with its heritage. As a teenager growing up on the south coast, all I could think of was getting out of a part of England that was increasingly becoming nothing more than a retirement zone for wealthy pensioners. But absence makes the heart grow fonder, and on return visits I began to take more and more interest in the traditions and customs of this quite beautiful county. Here was a distinctive rural culture, with a peculiar dialect, a rich folklore and distinguishing local cuisine. And taking a modest place among the pantheon of Sussex's customary idioms was the humble trug. I'd seen trugs as a boy, on fruit farms and slung from the arms of the older gardeners at the civic allotments, but I hadn't realised that they were particular to Sussex.

Although the first truggers are mentioned in the sixteenth century, more detailed records show that a man called Thomas Smith began making trugs in the late eighteenth century in a place called Herstmonceux – just across the marshes from where I'd lived in my youth. At this point of revelation – while reading a craft book in a library in London – a fountain of Sussex pride welled up inside me, and I vowed to procure my own authentic trug. This I eventually did, some time later, at an antiques shop in Bexhill-on-Sea. The proprietors saw me coming, detecting my initial glee at spotting it and my intense interest as I turned it round in my hands inspecting the build for authenticity. 'Forty pounds', said the wily old lady behind the counter. 'Forty pounds?!' I protested. But, of course, I *had* to have it. It was a rite of passage.

Back at my cottage in Wiltshire, I found excuses to use my new acquisition, proudly making repeated journeys transporting vegetables and fruit from the garden to the kitchen. It was clearly old when I bought it, a slight touch of woodworm and various stains were testament to its working pedigree. Since then I've given it a hard time, loading it with muddy potatoes, sodden leeks, soft fruit and all manner of earthy garden materials. Once again, few receptacles surpassed this well-made basket for strength, durability and lightness, all derived from the *cræft* in its making. Many of the techniques and tools were similar to those used to make a spale basket, but the trug was remarkably different in style.

Like the spale basket, the frame would have been made from a rod of hazel, but in this case, split down the middle and shaved down for evenness. This would be steamed and then manipulated around a mould before being fastened into a loop using copper clout nails. A second, smaller loop would make up the handle and this would be fastened at right angles to the first. Again, like the spale basket, the main body of the trug was made with split timber, but in this instance chestnut, a popular wood in Sussex, or white willow were used, both sourced from coppiced poles some six to eight years old. The form of the resulting strips was different. The trug was made using much broader strips, or rather boards of cleft wood where a width of an eighth of an inch was most desirable. The better you were at splitting out the boards, the less work was required to shave them down with a drawknife, producing less waste material. The finished boards would be saturated and set on a 'brake', a kind of post or frame with wooden pegs fixed in such a way as to bend the soaked ends upwards at each end. Once dried, they are attached to the frame, again, using copper clout nails. Here comes the main difference from the spale basket; these boards aren't woven in the same way as the spelks and chissies of the spale, but rather laid longitudinally and with a slight overlap. The sturdiest would be used as a centreboard, then came the 'seconds' on each side, and then side boards were used to make up the body so that the finished article looked like an inverted clinker-built boat.

I THINK THE MOST awestruck I have ever been by a craft object was in Oslo, at the Viking Ship Museum. Standing at the prow of the Oseberg ship, I was transfixed by the sleek, organic lines of the clinker-built hull as they dropped effortlessly away from the decorated stem post, swelling outwards, caressing the belly of the midship. In the dim hush of the exhibition hall I could almost imagine the creaking of the rigging, the trickle of water from the oar slice, and feel the vessel gliding through the shallows, cutting the mists as it stalked its quarry. Excavated in 1904, the Oseberg ship dates from the mid-ninth century and was recovered from a burial mound on the west bank of Oslofjord. It shares space in the museum with two other ships from a similar period, the Gokstad and Tune, which were both recovered from burial mounds in a similar location and derive from a similar period. It's their style of construction that's so impressive. Known from vessels dating from the third to the thirteenth centuries, it's the period represented by the Oseberg, Gokstad and Tune ships that marks a high point in the technological abilities of the Viking shipbuilders. To the Norwegian examples can be added the five Skuldelev ships recovered from the waterway of Peberrenden in Denmark, which owe their remarkable survival to the fact that they had been hulked (deliberately wrecked) to create a barrage across the fjord to protect the royal town at Roskilde from seaborne attack. The Norwegian and Danish examples cover a period of around three hundred years, an era that has come to be known as the Viking Age.

Not all Viking ships are the same; regional styles would have prevailed and different boats would have been built for different functions. The most renowned, the warship, was long and thin, with decking and a full complement of oars. These were designed to carry men and high-value, low-volume goods such as gold and silver. Shorter and wider boats would have been built for cargo and even smaller boats built for fishing. While the Vikings were most famous for their brutal raiding parties, their vast empire – one that stretched across Iceland, Ireland, the Scottish fringe, northern England, parts of Wales, Scandinavia, beyond the Baltic and into modern-day Russia – could only be sustained through a first-class merchant navy. The Skuldelev examples, because of the nature of their deposition, are more representative of the various styles of ship constructed in the age, but despite the differences in size, shape and function, the principles of construction are broadly the same.

The fundamental principle, and something it would have been unthinkable not to have done at the time, was to always, always, always work with the grain of the wood. So the building of a Viking ship began in the forest and in the selection of timber to match the job that was required of it. Tall forest oak would be sought out for the planking, or 'strakes', mast and keel. Allowed to grow naturally in open spaces, oak will develop large muscular and curvaceous boughs. These would be used in the carving of side-framing timbers and the 'knees', the angular blocks that brace the cross-beams of the frame. Forked timbers would be sourced to make a keelson – a footplate that sat on the keel in the middle of the boat, designed to host the mast. The stem and sternposts would be carved out of a single trunk of wood and, remarkably, the first two to three feet of each strake would be simulated through the carving of 'wings'. The logic behind this was that each strake could be more securely fastened to these simulated strake ends than they could to the post itself.

Timber was always worked green and never in a seasoned condition. For the strakes, trunks were radially split with the expectation that some twenty planks should be achieved from a trunk of around three feet diameter. Iron or seasoned hardwood wedges would be used for this process, and the strakes, wedge-shaped in profile, were then ideal for overlapping with each other. The accuracy of the split and the standard of the finished strake were critical to the success of the vessel. If the planks were too thick, then the overall weight of the ship would cause it to sit lower in the water, creating greater draught friction. So it was essential to get the balance right between strength and lightness, and some idea of where that balance lay can be seen in the lower planks of the Gokstad ship which has a thickness of around one inch. Probably the best account of Viking shipbuilding comes from the scenes in the Bayeux Tapestry where William of Normandy's men are depicted constructing a fleet to invade England. Descendants of settling Norsemen, the Normans had Viking blood in their veins, and in the tapestry is a pictorial illustration of the processes and tools involved, from the felling of trees through to the sailing of the final vessel. The axe dominates as the main tool, with four different types represented. Augers and chisels are also depicted alongside adzes, an axe-like tool but with the blade set on a horizontal plane. Close examination of the tool marks on all the surviving ships confirms what the Bayeux Tapestry so graphically demonstrates – Vikings were truly masters of the axe.

What marks out Viking ships as so impressive is the way they are constructed. Boat construction today is lazy by comparison. You build a frame and then force machine-sawn timbers to bend to the shape of the frame, meaning there is already a tension between skin and frame that could wrest them apart. Working with the grain of the split strakes, the Nordic shipbuilders established an inherent strength in the hull before inserting the frame. So the keel (the spine that runs down the centre of the hull – a bit like the centreboard in the trug) would be attached to the pre-carved stem and sternpost and then the garboards – the first tier of strakes – would be fastened onto the keel. Depending on which part of the Viking Empire the ship was being built in, the strakes would be caulked – waterproofed – using either moss or tarred animal hair. Different methods between east and west were used to fasten strake on strake. In the west, the method consisted of iron nails driven through both strakes before passing through a 'rove' (a type of washer), against which the tip of the nail would be hammered flat. In the east, rather more ingeniously, treenails (pegs) were hammered through pre-drilled holes and a small wedge driven into the split end of the peg to open it out and secure it in place. Under constant submersion, the peg would swell and further fix the two swollen strakes fast. Symmetrical strakes would then be attached either side of the garboards, overlapping to create the clinker style of boat. When the lower hull had been completed, it was only then that the supporting floor timbers were inserted, and as the strakes were built up on each side, further internal side-framing timbers were added.

There are some truly crafty design features in a Viking ship. First, the hull was symmetrical at both ends so it could be rowed forwards or backwards. This was useful for travelling up and down river, and for beaching without having to turn the vessel around. It also meant that a speedy getaway could be planned. Purposely, thicker and broader strakes would be used at the waterline, reflecting a need to protect it from the inevitable expansion and contraction which comes with the part of the boat that experienced frequent changes in saturation levels. The uppermost strakes were also more substantial to accommodate the oar ports and access in and out of the boat – you needed a good strong foothold from which to launch yourself ferociously upon your enemy. Finally, to the side and floor-framing would have been fixed cross-beams to support decking or benches.

Bulkheads would have further reinforced the joint between stem and sternposts and strakes, and a keelson and mast step would have been inserted to carry the mast.

The Viking longship is such a remarkable craft achievement in part because the level of sophistication in the build was derived from such a simple set of tools. Most important is the adherence to the golden rule of working *with* the wood, capitalising on the inherent strength of timber split along the line of the grain. It's this modest principle that connects my humble trug with the great Viking longship; the *cræft* of gaining maximum strength from minimum thickness, of taking a single piece of material that on its own might not be of remarkable robustness but when laid – either longitudinally or in a weave – against itself, creates something of astounding strength, lightness and beauty. There is a quiet humility to these qualities in the trug and the spale basket, but for the Viking longship, built like a giant basket, it had no internal frame to weigh it down, and as a consequence could not only travel in the shallowest of rivers, penetrating deep into the heart of enemy territory, but could also roll with the punches of the open sea.

It's a craft that relies on building something relative to the materials employed, rather than to some kind of formal blueprint. It's a moment of true craft creation. Rather than recreating from a preconceived plan of constraining measurements, it's about allowing the materials to speak for themselves, to answer back, to tell you what the natural shape must be if you're to make maximum use of their natural properties. For me, the Vikings took this philosophy to its extreme. For a period of three hundred years they dominated the waters of northern Europe. It's now accepted that they were much more than raiders intent on pillage and destruction. Contemporary scholarship acknowledges the fundamental part they played in the development of medieval Europe. They settled areas that it would be unthinkable to settle today without access to the most advanced technology. They forged kingdoms and allegiances of lasting importance and set up a trading network that spread from the western Mediterranean in an arc up through the west of Europe and Britain, through Scandinavia, the Baltic, Russia and south into the Byzantine Empire. Arguably, among their greatest maritime achievements was the discovery of America – and they couldn't have done it without the *cræft* evident in my antique trug and Owen Jones's spale basket.

POSTSCRIPT

CRÆFT AND CONTEMPLATION

THE EIGHTEENTH-CENTURY GERMAN philosopher Immanuel Kant, in his *Critique of Judgement,* differentiated between free and dependent beauty, and this is a useful distinction when considering the future of British craft. In the world of art, free beauty can be tainted by a reliance on the functional: form and appearance should always be appreciated as part of a pure aesthetic. But I would judge the forging of a good billhook, not on how pleasing aesthetically it is to the eye but on how close its form is to that of other billhooks I've used. Its attraction – and therefore its beauty – is dependent on its ability to function as a billhook.

We've become pretty adept, in the western world, at enjoying and celebrating the pure aesthetics of crafted objects, no doubt a legacy of the Arts and Crafts movement. But we increasingly struggle with dependent beauty because we don't know how to place or use functioning crafted goods. We don't know what to measure that beauty against. What is it dependent on? We struggle with the true value of a woollen blanket because our central heating never allows us to get cold enough. We can't appreciate the ergonomics of a good scythe because we don't need to make hay. We aren't aware of the cooling properties of a stoneware storage jar because we've only ever known life with a fridge. We care less about the workmanship invested in our shoes because when they start to fall apart we just buy another pair. If the 1970s craft revival movement in Britain was about saying, 'Hey, we're losing something – the last generation of makers is dying out!', then the meta-narrative of the current crop of craft-orientated writing is about disconnection: we have become detached from making, and it isn't a good state for us to be in. It's unhealthy when we are disconnected from making.

At its deepest root, I think this disconnection is derived from our illiteracy of power. Beyond getting a bit puffed out after hefting a few boxes up the stairs, we don't, in society as a whole, really know what energy is all about. It's too macro for us to comprehend. We've become too used to electricity or gas on tap – flicking a switch and using as much as we can afford – facilitated by increased automotive and mechanical complexity. This is the age of the leaf blower, the electric window and the battery-powered pepper grinder (with a built-in light). The people who argue that the price of wind and tidal energy will have to come down before we can further invest in them are so enslaved to an economic model founded on petrochemicals that they completely miss the point: we've never had power so cheap and, arguably, will not have again. Our over-reliance on fossil fuels, stemming from the Industrial Revolution, has sold us down a dead-end street. And we can't afford to back up.

Craft has, and always will, enjoy buoyancy among the luxury markets, for those who can appreciate it and for those who are simply buying a price tag. But for the everyday, the cost is prohibitive. Take drystone walling. Even in the Cotswold region of England – where money is in no short supply, delivery is cheap and raw materials are widely available – wire fencing still proliferates over dilapidated stretches of ruined walls. Why, when a drystone wall is clearly a superior form of land boundary? Because we haven't found a machine to do the job. Because it takes a hand–eye co-ordination that only the most complex machine we have at our disposal – our own body – is capable of. But we can no longer afford to pay for this human source of power.

This may change, though. Fuel costs, in real terms, are rising. We're on the brink of an energy crisis and there is an emerging oil shortage. More to the point, planet Earth is roasting and, if it isn't too late, we may just be able to pull it back from disaster. If we do, it will be because it's cheaper to go back to using people, rather than machines, to do the work. Rake up your leaves. Grind your own pepper. Use your own arm to wind your car window up and down. Use your legs, in the first place, to get you from A to B. Perhaps, when we once again start to use our own bodies – our own kinaesthetic sensibility – to support our existence we may regain a literacy of power and rediscover the knowledge of *cræft*. Our problem with British craft is that we have over-fetishised the point at which a crafted object is put together. We've conflated craft with skill and design with art when, following the German definition of *Kraft*, it should be about more than just making. It is the power, the force, the knowledge and the wisdom behind making – the *cræft* behind it.

I OFTEN LIKE TO say that all traditional craft ways lead to the farmhouse table. Whether a fruit-picker's basket, a blacksmith's ploughshare, a dairymaid's cream dish, a shepherd's crook or a labourer's boots, such items were part of systems of food production that fall under the broader definition of farming. Even when thatching with cereal straw or weaving with wool, agriculture has come into the equation somewhere. This has, ultimately, been my route into traditional rural crafts. The baseline craft of farming, the foundation on which all rural crafts are built, is undoubtedly that of digging; understanding the soil, how it behaves, how it gives life, and the expense in energy needed to work it. Building on the tilling of the soil is the knowledge of your seeds; how they grow and when to harvest them; the knowledge of how to develop a grassland sward and how to nurture it so that it can support the livestock. But of all these crafts that punctuate the agricultural year and pepper the rural world, what I've always found so intelligent about them is their resourcefulness and their distinctly local character. Crafts have always been determined by the immediate environment, and indexed to the resources of the natural world. That's why they've been around for so long. I like to consider the true trajectory of craft production and use as one that runs like so:

tended landscape → sustainable production of raw materials → intelligently processed → beautifully made → fit for purpose → fondly used → ingeniously reused → considerately discarded → given back to the earth

Herein lies the true *cræft* – the power, the knowledge and the skill – in the rural crafts of old. Making hay is a classic example of this. Popular craft books present the traditional making of the scythe and the hay rake as 'craft' objects. But do we really comprehend them as stand-alone artefacts? Should they be seen in isolation from the *cræft* of haymaking, the practice that makes them so intelligent in the first place? It's clever to make a scythe, it's even cleverer to use it effectively.

I get angry over the lack of basketry in our lives. Up against its closest competitors – cardboard and plastic – it emerges the outstanding winner. By their nature, the raw materials for basket making must come from self-sustaining sources, the year-on-year regrowth from coppiced stools of hazel, willow, alder or any other species that sends up sucker shoots from a harvested stool. And yet these materials are mechanically thrashed out of the hedges in our landscapes, or clipped and trimmed from the shrubs in our gardens, deposited in the green-waste bin and carted off to the local recycling centre. Do we, therefore, really need quite so much deforestation-produced cardboard or quite as many plastic crates, the latter a casual by-product of the petrochemicals industry? Both, like baskets, don't last for ever.

But when a basket's working life is over, it can be left to rot, to be given back to the earth and to be replaced at no cost to the environment. We just need to give someone the time to make it for us. We need to embrace the cyclic economy. We make for profit and not for use. We are enslaved to growth economics.

What I've learned from so many crafts is that they are part of a trajectory of production and use, they are part of a cycle of life. Sometimes, as in the case of transhumance – the agricultural system that has produced meat, milk and fleece for millennia – the cycle is yearly, wedded to the life-giving properties of the passing of the seasons and the symbiosis of upland and lowland environments. But at other times the cycle is longer. The lifespan of a queen bee determines the pattern of harvesting and propagation in the skep beekeeper's domain. The durability of water reed or wheat straw determines the number of times a thatcher will return to rethatch a cottage in his own lifetime. The hedge layer will find himself standing in the same place at the same time of year, on a cycle of seven or eight years, laying the larger suckers, closing up the gaps and trimming out the brash.

Writing this book has also brought home to me what I call the 'deep time signatures' of so many of our crafts. It's part of the basic human condition to use a stick, and the point at which we started doing it routinely and purposefully marks an upward turn in the evolutionary journey of mankind. In Britain people still made a living using sticks well into the twentieth century. But we also see this continuity in fairly sophisticated technologies like weaving, a craft that goes back over four thousand years, and one that will be with us for ever. It's often dangerous to draw developmental arcs over such long time frames, to oversimplify very complex processes of cultural transmission. But even if there were periods of rupture and technological and evolutionary discontinuity – as in pottery production from the Romano-British through the Anglo-Saxon and into the medieval period – we should take heart that certain crafts are reborn and readapted when needed. If we ever find ourselves running short on ceramic crockery and kitchenware because of some apocalyptic collapse in the global network of exchange, I know that, like the Anglo-Saxons, I need to use a chaff temper when I'm mixing the short clay sourced from the bottom of the garden.

These deep time signatures also serve as a tacit reminder of the human condition: that we are makers, and that we have always lived in a world of making. It defines us, we need it, it's good for our health, and it makes us 'better'. Having given us the factory floor and mechanised production, the Victorians were quick to realise this. There is no finer illustration of how they felt craft could be used as a tool in moral reform than the establishment of Toynbee Hall in East London, the Edinburgh Social Union and the Kyrle Society institutions in Birmingham, Leicester and Glasgow. These ideas were to continue into the twentieth century, not least when, in the aftermath of two world wars, a generation of young men needed a remedy that cured their souls as well as their broken bodies. As part of this programme to rehabilitate blind, invalided and mentally crushed ex-servicemen, my grandfather, with repeated infections in his war wound, found himself stitching a rug as he convalesced in a hospital in south-west London in the 1950s. My grandmother, now in her eighties, has her husband's handicraft serving as a bedspread. Since his passing, it continues to provide some remedy for her loss.

I think John Ruskin was right. Factory manufacture robs us of a special something: contemplation. Not of life, of love, of the big 'Who am I? Where am I? What am I doing?' But, in the case of the drystone waller who holds apparently identical stones in each hand, the simple cognitive contemplation between one stone and the other. Which to use? How to work it? Where to strike it? – thoughts that exercise the mind in silence and solitude. This, I believe, is what we truly lack in today's society. Crafts are a vehicle through which we can think, through which we can contemplate, and through which we can *be*.

Cræft is a form of intelligence, an ingenuity that can shift in accordance with a changing world. What has seemed intelligent for the best part of 150 years – factory production, mass manufacture, conspicuous consumption and waste – now no longer feels all that intelligent in a world of diminishing resources and increasing environmental instability. So a new *cræfty*-ness is required, a rethinking of what it means to be powerful, resourceful and knowledgeable through the medium of making, the medium that defines us as human beings. This is unashamedly an idealist's book – I make no bones about it. To be *cræfty* is all about resourceful living and about going back to the basics: a mindful life achieved through beautiful simplicity.

ACKNOWLEDGEMENTS

My greatest debt of gratitude is owed to the very many crafts-people who have been so generous with their time, enthusiasm and support over the years. Without their input, I could never have produced this book. Walter Donohue has been a constant source of warm encouragement – his hand gently steadying the tiller of our collective Viking longship. Patrick Walsh has been brilliant from the start, and James Holland an endless source of enthusiasm. The team at Faber & Faber have been wonderful, in particular Samantha Matthews, and I am grateful for the keen eyes of Anna Swan and Sarah Barlow. Harry Brockway's engravings are amazing – when I saw the first sketches, I couldn't believe my luck. My family have been a huge support to me over the years and I thank my parents who, during my childhood, gifted me so many opportunities to play with sticks, stones, mud, wood, etc. During those formative years, my brother, Thomas, was always at my side and I now savour the all-too-rare moments we get to relive those halcyon days and engage in a craft pursuit together. In more recent years it has been my good friend Peter Ginn who has been at my side, one of the most patient, intelligent and caring people I have had the pleasure to work with.

I owe a lot to the Institute of Archaeology (University College London), the Heritage Crafts Association and the English landscape. But I reserve the most special thanks for my dear wife, Libby, and our magical children, Hazy and Pip.